Free Space Optics Technologies in B5G and 6G Era - Recent Advances, New Perspectives and Applications

Edited by Jupeng Ding, Jian Song, Kejun Jia and Penghua Mu

Published in London, United Kingdom

Free Space Optics Technologies in B5G and 6G Era - Recent Advances, New Perspectives and Applications
http://dx.doi.org/10.5772/intechopen.110928
Edited by Jupeng Ding, Jian Song, Kejun Jia and Penghua Mu

Contributors
Ahmad Adnan Qidan, Barzan Yosuf, Bastien Bechadergue, Bela Berde, Changiz Ghobadi, Chih-Lin I, Dominique Chiaroni, Elham Sarbazi, Fathi Abdeldayem, Harald Haas, Hardik Soni, Hossein Kazemi, Ian H. White, Jaafar Elmirghani, Javad Nourinia, Jian Song, Jintao Wang, Jose Tabu, Jupeng Ding, Majid Safari, Mazen Abdullatif, Meysam Jalali Asdabadi, Michael Crisp, Micheline Perrufel, Mohammad Dehghani Soltani, Mohammadreza Rohaninezhad, Nikola Serafimovski, Raman Jee, Ravinder Singh, Richard Penty, Rui Chen, Sanaa Mohamed, Seyed Ramin Emadian, Shenjie Huang, Somnath Chandra, Taisir E.H. El-Gorashi, Wajahat Ali, Yi Liu

First published in London, United Kingdom, 2024 by IntechOpen
IntechOpen is the global imprint of INTECHOPEN LIMITED, registered in England and Wales, registration number: 11086078, 167-169 Great Portland Street, London, W1W 5PF, United Kingdom

British Library Cataloguing-in-Publication Data
A catalogue record for this book is available from the British Library

Additional hard and PDF copies can be obtained from orders@intechopen.com

Free Space Optics Technologies in B5G and 6G Era - Recent Advances, New Perspectives and Applications
Edited by Jupeng Ding, Jian Song, Kejun Jia and Penghua Mu
p. cm.
Print ISBN 978-0-85466-389-7
Online ISBN 978-0-85466-388-0
eBook (PDF) ISBN 978-0-85466-390-3

For EU product safety concerns:
IN TECH d.o.o., Prolaz Marije Krucifikse Kozulić 3, 51000 Rijeka, Croatia,
info@intechopen.com or visit our website at intechopen.com.

Meet the editors

Jupen Ding received an MSc and Ph.D. from the Beijing University of Posts and Telecommunications, China. In 2013, he joined the Key Laboratory of Wireless-Optical Communications, Chinese Academy of Sciences, School of Information Science and Technology, University of Science and Technology of China. In 2017, he worked for one of the world's top 500 enterprises. Then, he joined Xinjiang University, in China. He has published more than fifty journal and conference papers and holds more than twenty national and international patents. He is a senior member of the Chinese Institute of Electronics, the Chinese Optical Society, and the China Institute of Communications.

Jian Song received a BEng and Ph.D. from the Electronic Engineering Department, Tsinghua University, China, in 1990 and 1995, respectively. He was a postdoctorate at the Chinese University of Hong Kong, China, and the University of Waterloo, Ontario, in 1996 and 1997, respectively. He then worked in the United States for seven years and joined Tsinghua University in 2005 as a full professor. He is now the director of the DTV Technology R&D Center, one of the major technical contributors to the Chinese DTV standard. Dr. Song has served as an associate editor of *IEEE Transactions on Broadcasting* since 2010 and is the general chair of IEEE SmartGridComm 2019 and IEEE BMSB 2014. He was elected the vice chairman of WP 6A of the International Telecommunication Union (ITU) in 2012 and is the founding editor-in-chief of two ITU-related academic journals, *ICT Discoveries* and *Intelligent and Converged Networks*. Dr. Song has published more than 300 peer-reviewed papers. He also holds two US patents and more than eighty Chinese patents. He is the co-author of several books and book chapters.

Kejun Jia received his MS in Communication and Information Systems from Xi'an University of Technology, China, and his Ph.D. in Communication and Information Systems from Southwest Jiaotong University, China, in 2007 and 2018, respectively. He is a professor at the School of Computer and Communication, Lanzhou University of Technology, China. His research interests include indoor visible light communications, multi-carrier systems, and next-generation wireless communications.

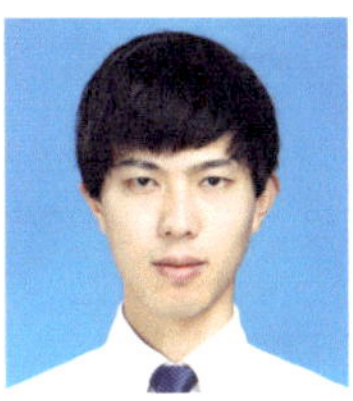

Penghua Mu obtained a BS in Communication Engineering and a Ph.D. in Optoelectronics from Southwest Jiaotong University, China, in 2009 and 2016, respectively. He joined the School of Physics and Electronics Information, at Yantai University, China, in 2018, where he is currently an associate professor. He has published more than 100 research papers. His current research interests include nonlinear dynamics of semiconductor lasers, random number generation, and chaotic security communication.

Contents

Preface

Conventional radio frequency (RF) bands are becoming overcrowded and are unable to accommodate emerging services such as augmented reality, mixed reality, 8K video, and self-driving vehicles. To address the growing capacity demands and performance challenges of networks, it is essential to explore and integrate novel free space optical (FSO) techniques and wireless communication methods to leverage their complementary characteristics.

FSO techniques offer advantages such as being license-free, inherently secure, and having relatively low infrastructure costs. In contrast, wireless communication techniques provide robustness, widespread coverage, and higher maturity. To design and develop future 6G networks, researchers from both academia and industry are actively contributing and proposing innovative schemes. The chapters in this book examine relevant topics and present analyses from the perspective of FSO and wireless communication technologies.

The need for disruptive transmission approaches in emerging application scenarios has led to a convergence of FSO technologies with B5G and 6G transmission services. This book explores the practical considerations of this integration and investigates how FSO technologies can address challenges in B5G and 6G environments. In addition, it provides foundational understanding of recent advances and perspectives in the field.

Jupeng Ding
School of Information and Electronic Engineering,
Shandong Technology and Business University,
Yantai, China

Key Laboratory of Signal Detection and Processing,
School of Computer Science and Technology (School of Cyberspace Security),
Xinjiang University,
Urumqi, China

Jian Song
Department of Electronic Engineering,
Beijing National Research Center for Information Science and Technology,
Tsinghua University,
Beijing, China

Penghua Mu
Associate Professor,
School of Physics and Electronics Information,
Yantai University,
Yantai, China

Kejun Jia
Professor,
School of Computer and Communication,
Lanzhou University of Technology,
Lanzhou, China

Section 1

Introduction and Overview

Chapter 1

Introductory Chapter: Novel Free Space Optical and Wireless Communications for Future-Oriented 6G Networks

Jupeng Ding

1. Introduction

According to the commercial forecasting, the global mobile traffic is expected to be more than 5000 EB/month, and the predictable number of connected devices on the Internet will be up to 500 billion devices by 2030. Therefore, the conventional radio frequency (RF) networks are already overcrowded and could not satisfy the quality of service of 6G network independently [1–4]. For tackling the above challenges, considering the available radio spectrum is getting nearly saturated, diverse free space optical (FSO) techniques have been proposed and are becoming one promising solution to conventional RF network paradigm [5–7]. The charming and unique features of FSO include hundreds of terahertz of unlicensed optical spectrum, high security, electromagnetic interference-free operation, and simple implementation. At the same time, considering the disadvantages such as limited coverage and easy to be blocked for the FSO network alone, the novel wireless communication techniques must be consistently investigated to achieve reliable and seamless coverage simultaneously.

2. Novel free-space optical communications for future-oriented 6G networks

For further enhancing the transmission capacity performance of the FSO system for the evolving 5G and future-oriented 6G wireless networks, the wavelength division multiplexing (WDM) technique is introduced to FSO system in the presence of wireless and optical nonlinearities. Moreover, for typical indoor application scenarios, the researchers are actively exploring the superior performance features of laser-based wireless communications, compared to the conventional light-emitting diodes-based wireless communications with limited several tens MHz modulation bandwidth. As for outdoor lane scenario, the fundamental channel characteristic is compared between the distinct non-Lambertian LED beam and Lambertian LED beam configurations for 6G wireless networks, for evaluating the effect of spatial optical beams on vehicle FSO performance.

3. Novel wireless communications for future-oriented 6G networks

Similarly, the antenna systems and antenna mutual coupling are vital to the performance improvement of future-oriented 6G networks as well. Mutual coupling in the context of antennas presents a substantial and pressing concern within the realm of 5G and 6G wireless communication systems. In particular, one innovative method is proposed by the researcher in this book to minimize mutual coupling between antennas without permanent alteration of the structure. In addition, advancements in antenna systems are systematically reviewed for beyond 5G (B5G) and anticipated 6G. The concerned antenna aspects include microstrip antennas, metamaterial-based designs, reconfigurable antennas, phased array antennas, and lens antennas.

Acknowledgements

This work is supported by the National Natural Science Foundation of China (Grants No. 62061043), the Natural Science Foundation of the Xinjiang Uygur Autonomous Region (Grants No. 2019D01C020), and the Tianshan Cedar Project of Xinjiang Uygur Autonomous Region (Grants No. 202101528).

Conflict of interest

The authors declare no conflict of interest.

Author details

Jupeng Ding
Key Laboratory of Signal Detection and Processing in Xinjiang Uygur Autonomous Region, School of Computer Science and Technology (School of Cyberspace Security), Xinjiang University, Urumqi, Xinjiang, China

*Address all correspondence to: jupeng7778@163.com

References

[1] Chowdhury MZ, Hasan MK, Shahjalal M, Hossan MT, Jang YM. Optical wireless hybrid networks: Trends, opportunities, challenges, and research directions. IEEE Communications Surveys & Tutorials. 2020;**22**:930-966. DOI: 10.1109/COMST.2020.2966855

[2] Pham T, Atsushi K, Keizo I, Toshimasa U, Naokatsu. Hybrid optical wireless- mmWave: Ultra-high-speed indoor communications for beyond 5G. In: IEEE INFOCOM 2019 - IEEE Conference on Computer Communications Workshops (INFOCOM WKSHPS). Paris, France: IEEE; 2019. pp. 1003-1004. DOI: 10.1109/INFCOMW.2019.8845283

[3] Huang C, Hu S, Alexandropoulos GC, Zappone A, Yuen C, Zhang R, et al. Holographic MIMO surfaces for 6G wireless networks: Opportunities, challenges, and trends. IEEE Wireless Communications. 2020;**27**:118-125. DOI: 10.1109/MWC.001.1900534

[4] Khan M, Chakareski J. Neighbor discovery in a free-space-optical UAV network. In: 2019 IEEE Global Communications Conference (GLOBECOM). Waikoloa, HI, USA: IEEE; 2019. pp. 1-6. DOI: 10.1109/GLOBECOM38437.2019.9014063

[5] Georlette V, Bette S, Brohez S, Pérez-Jiménez R, Point N, Moeyaert V. Outdoor visible light Communication Channel Modeling under smoke conditions and analogy with fog conditions. Optics. 2020;**1**(3):259-281. DOI: 10.3390/opt1030020

[6] Moreno I, Sun C. Modeling the radiation pattern of LEDs. Optics Express. 2008;**16**(3):1808-1819. DOI: 10.1364/OE.16.001808

[7] Ding J, Chih I, Xu Z. Indoor optical wireless channel characteristics with distinct source radiation patterns. IEEE Photonics Journal. 2016;**8**(1):1-15. DOI: 10.1109/JPHOT.2015.250842

Section 2

Free Space Optical Communications Systems for Future-Oriented 6G Networks

Chapter 2

Indoor Laser-Based Wireless Communications

Ahmad Adnan Qidan, Mohammad Dehghani Soltani, Wajahat Ali, Rui Chen, Shenjie Huang, Barzan Yosuf, Sanaa Mohamed, Ravinder Singh, Yi Liu, Hossein Kazemi, Elham Sarbazi, Bela Berde, Dominique Chiaroni, Bastien Bechadergue, Fathi Abdeldayem, Hardik Soni, Jose Tabu, Micheline Perrufel, Nikola Serafimovski, Taisir E.H. El-Gorashi, Michael Crisp, Richard Penty, Ian H. White, Harald Haas, Majid Safari and Jaafar Elmirghani

Abstract

Optical wireless communication (OWC) is a strong candidate in the sixth generation (6G) of wireless communications as it can support data transmission at high communication speeds, low power consumption, high security, and high reliability. Near-infrared lasers, specifically vertical cavity surface emitting lasers (VCSELs), exhibit larger modulation bandwidth compared to light-emitting diodes (LEDs), and can provide aggregate data rates ranging up to Terabit per second (Tbps). This chapter proposes innovative optical transmitter and receiver designs for indoor laser-based wireless communications, aiming to provide seamless coverage in high-speed multi-user scenarios. The proposed optical transmitter is based on a 5×5 VCSEL array where each VCSEL can provide 10 Gigabit per second (Gbps) in a uniform coverage area of 20 cm x 20 cm, achieving a total of 250 Gbps in an area of 1 m^2 at 3 m transmission distance. Moreover, two wide-FOV receiver designs that employ photodetector (PD) arrays in along with imaging or non-imaging optics are introduced to support such an ultra-high-speed optical link. The new design of OWC requires potential techniques to address various networking challenges, including cell formation, interference management, resource allocation, and backhauling. To this end, potential solutions for these challenges are discussed to help establish an end-to-end OWC infrastructure for practical deployment.

Keywords: 5G wireless networks, 6G wireless networks, optical wireless communications, LEDs, infrared lasers, networking

1. Introduction

The number of wireless communication devices is increasing rapidly, with the goal of providing easy internet access everywhere, unobstructed broadband services, reliable monitoring of users' health conditions and environmental conditions, etc. In the next generation of wireless networks, aggregate data rate in a range of Tbps must be achieved to support the unprecedented growth of data traffic. Optical wireless communication (OWC) can offload heavy traffic loads from current radio frequency (RF) wireless networks as it operates over the expansive and license-free optical spectrum.

Light fidelity (LiFi) is a version of OWC technology that uses light-emitting diodes (LEDs) as optical transmitters to deliver high-speed data communication in a similar manner to WiFi. Basically, LiFi is a bidirectional and fully networked wireless communication usually deployed in indoor environments to provide both illumination and communication [1, 2]. Several types of LEDs preferred for use in this context are listed as follows:

1. **Phosphor converted LED (PC-LED):** A single blue Indium Gallium Nitride (InGaN) LED chip is used to pump a Yttrium Aluminum Garnet (YAG) phosphor coating. The phosphor converts part of the blue light to green, yellow and red portions of the spectrum while the other part of the blue light is leaked out, and the mixture of which produces white light [3].

2. **Multi-chip LED:** Such LEDs generally combine three or more sources emitting different colors, typically red, yellow, green and blue (RYGB). Compared to the PC-LED, the multi-chip LEDs offer higher aggregate data rates with the utilization of wavelength division multiplexing (WDM).

3. **Micro-LED (μ-LED):** This is an advanced form of LEDs that is able to achieve modulation bandwidths in a range of hundreds of MHz [4], and therefore provides high data rates per transceiver chain. The μ-LEDs have a small size ($\leq 100\mu$m) when compared to other types of LEDs. As a result, they rely on a relatively large number of sources to provide sufficient illumination for indoor lighting purpose.

4. **Organic LED (OLED):** It is a popular technology for flat panel displays, in which the light is generated from organic thin films sandwiched between an anode and a metal cathode. The typical modulation bandwidth of OLEDs is around hundreds of kHz [5].

5. **Quantum Dot LED (QLED):** This type of LED uses quantum dots as light-emitting material. With different materials and processes, QLED has shown very impressive results so far for bright display panels and can offer modulation bandwidth up to MHz [6, 7]. However, this constrained bandwidth renders them less suitable as optical transmitters for multi-Gbit/s communication.

On the other hand, laser diodes, especially vertical cavity surface emitting lasers (VCSELs), provide larger modulation bandwidth compared to LEDs. VCSELs also exhibit other outstanding features such as high power conversion efficiency, low cost, long lifetime, and ease of manufacture, which make them the most promising candidates to fulfill the exceptional demands of the next generations of wireless networks. Near-infrared (NIR) VCSELs have shown the potential to support data transmission at

Tbps communication speed [8] with low complexity modulation and detection techniques compared to LED-based OWC, which requires advanced signal processing techniques to achieve Gbps aggregate data rates. However, two of the major challenges in laser diodes OWC are expanding the coverage area of the laser to support mobility considering the narrow beams, and ensuring that the transmit power can meet the eye safety regulations [9].

The aim of this chapter is to derive new components, transmitters and receivers, and define crucial networking techniques to unlock the full potential of OWC in 6G networks. We First propose an array of VCSELs as an access point that ensures uniform coverage up to a few square metres in order to provide seamless data transmission service in the real-world deployment of OWC. Then, two trade-offs in designing optical receivers are highlighted to enhance the field of view (FoV) and data rate. Finally, the chapter focuses on the limitations of laser-based OWC from the networking perspective and provides solutions to bring its implementation to reality.

2. 6G optical transmitters and receivers

2.1 Transmitters

Both LEDs and laser diodes can be used as the sources for OWC links. LEDs are the most common sources because they are widely deployed in indoor illumination. However, they suffer from relatively limited bandwidth, which restricts the achievable rates [1]. On the other hand, laser diodes (LDs) can provide much higher (>10 GHz) bandwidths and higher output power, making them more suitable for high-speed OWC links than LEDs [8]. Among different kinds of LDs, VCSEL can be a preferred choice for OWC, due to several key advantages. Firstly, VCSELs are much easier to manufacture and test than conventional edge-emitting lasers and they can be fabricated into two-dimensional arrays. Secondly, VCSELs require relatively simple optics for beam collimation due to their circular beam profile with low divergence. Thirdly, VCSELs have high wavelength uniformity and stability due to their short Fabry-Perot cavity.

Despite these advantages, VCSELs also exhibit certain drawbacks. Most evidently, commercially available high-bandwidth VCSELs are mainly multi-mode devices and exhibit a doughnut-shaped beam profile. Without proper optics, this particular type of beam profile gives non-uniform power and SNR distributions, which is problematic for OWC applications with wide coverage areas. Moreover, doughnut-shaped beam spots, when intended for full coverage in the far-field, create overlaps in the illuminated area, causing interferences between adjacent atto-cells.

To overcome these issues, a novel transmitter design is proposed which can be used with both multi-mode and single-mode VCSELs. Moreover, unlike many laser-based OWC designs in the literature which focus on single-user, high-speed point-to-point link with relatively long transmission range [10–12], the proposed transmitter design aims to provide a wide-area coverage for multiple users, offering a high tolerance for alignment and user mobility. Moreover, as each channel of the proposed transmitter generates a square atto-cell of 20 cm $\times$ 20 cm with high intensity uniformity (90%, verified through simulations and experiments [13]) and adjustable cell-to-cell distance, the inter-cell interference is minimized.

The schematic of the transmitter design is shown in **Figure 1**. In the proposed design, a micro lens array (MLA1) with the same pitch as the VCSEL array is used to

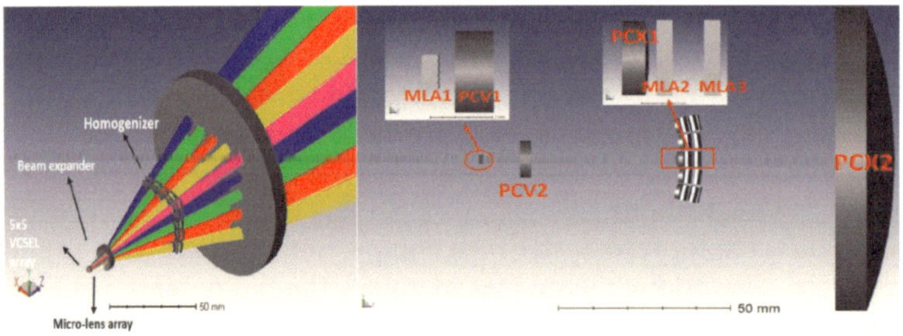

Figure 1.
System diagram (left) and lens arrangement (right) of the proposed VCSEL array [13].

collimate the transmitted beam. A wide beam-to-beam separation is achieved with the help of a pair of Plano-concave lenses (PCV1 & PCV2). After beam separation, each beam is coupled into an individual homogenizer formed by plano-convex lens (PCX), MLA2 and MLA3. The final image at 3 m is obtained by passing the output beams from the homogenizer through a PCX2, generating 20 cm $\times$ 20 cm coverage for each channel.

The designed transmitter can serve 25 users with 1m^2 coverage along with minimum inter-channel interference. Since each VCSEL in the array can achieve 10 Gbps, the transmitter can provide an aggregated capacity of 250 Gbps. The shape of the output beam depends on the pitch and radius of curvature (ROC) of each micro-lens array. Therefore, different types of micro-lens arrays can be used to vary the shape and size of the final beam spot on the receiver plane, making it easy to adjust the coverage area per picocell and the desired light intensity. By making use of off-the-shelf components a cost-effective transmitter can be built.

2.2 Receivers

The design of high-speed optical receivers is a crucial aspect in the development of indoor laser-based communication for 6G. The general structure of an optical wireless receiver must consist of optics to collect sufficient power, a detector, and the associated driving circuit, as illustrated in **Figure 2**. Developing high-speed, compact, and wide field-of-view (FOV) receivers, which can operate reliably in a mobile environment, presents inherent challenges due to two recognized trade-offs. The first trade-off is the area-bandwidth trade-off of the photodetector (PD). To support high data rates, small PDs with higher bandwidths are required. It is important to highlight that using a small-area PD, which has a high bandwidth, may not be preferable since it introduces less received power. One solution to address the area-bandwidth trade-off is to improve the receiver sensitivity by using avalanche photodiodes (APDs) or single-photon avalanche diodes (SPADs). However, the performance of these detectors is limited by their noise signal-dependency and limited dynamic ranges. Another effective and frequently employed approach to tackle this issue is employing optics in front of the PD to enhance the received power. However, this can lead to limitations on the receiver FOV as a result of the law of conservation of etendue, bringing the second trade-off known as the gain-FOV trade-off. Note that while narrow receiver FOVs are preferred in outdoor FSO systems with complex pointing and tracking subsystems capable of maintaining line-of-sight connectivity, indoor laser-based wireless systems typically require larger receiver FOVs to facilitate mobility support.

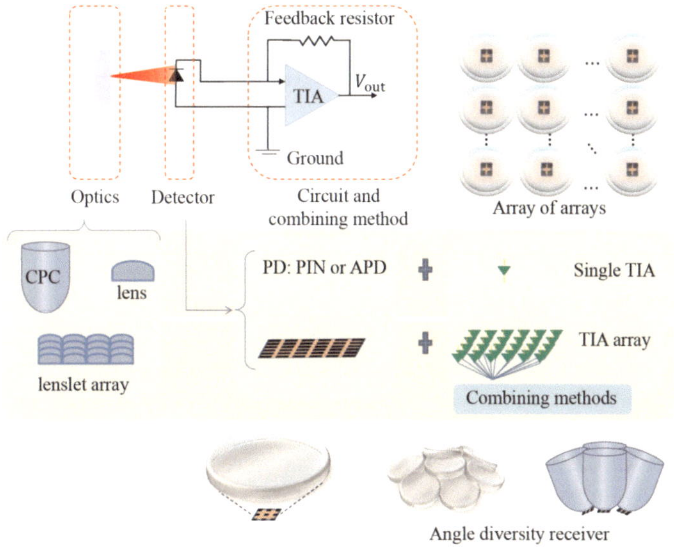

Figure 2.
General structure of optical wireless receivers.

The gain-FOV trade-off is caused by the use of optics, indicating that the optical gain is achieved at the expense of a reduced FOV. Depending on the types of optics, there are two broad categories of optical receivers: imaging receivers and non-imaging receivers. Imaging receivers use a lens or other optical element to focus incoming light onto a detector. Non-imaging receivers, on the other hand, rely on other mechanisms such as waveguides or fiber optic cables, to deliver light to the detector. Imaging receivers are typically more popular due to the availability of varieties of low-cost lenses with different sizes as well as lenslet arrays. Imaging receivers benefit from smaller heights, however, they have a smaller FOV compared to non-imaging receivers.

The combined effects of the aforementioned trade-offs lead to a new trade-off between the achievable data rate and receiver FOV, which is called rate-FOV trade-off [14]. This is due to the fact that attaining a higher data rate necessitates a broader bandwidth and an increased received power, both of which reduce the FOV. Several studies have investigated the enhancement of FOV of high-speed optical receivers. One approach is the utilization of an angle diversity receiver (ADR) which comprises several narrow-FOV receiver elements oriented in different directions [15], using non-imaging optics [16–18] or imaging optics [19], as illustrated in **Figure 2**. A wide-FOV receiver that incorporates fused fiber-optic tapers is proposed in [20], integrating hundreds of thousands of tapered optical fibers. The reported specifications include an overall FOV of 30°and an optical gain of 121.3. However, the compactness of receivers developed using these methods inevitably surpasses that of traditional receivers. An alternative, promising solution that effectively manages the rate-FOV trade-off, ensuring both high data rates and large FOVs while maintaining a compact design, is the use of PD arrays. To this end, several experimental works in the literature have demonstrated the use of an array of PDs in OWC. In [21], a broadband receiver featuring a large aperture is introduced, which can achieve a data rate of 1 Gbps at a FOV of 10°. In addition, an array of 8×8 PDs with a FOV of 6°is reported in [22], which achieves a data rate of 25 Gbps at a distance of 10 m.

We recently introduced two innovative wide-FOV receiver designs tailored for high-speed OWC systems, as presented in [14, 18]. In [14], a multi-element imaging receiver design based on an array of arrays is proposed, as depicted in **Figure 2**. In the proposed structure, every receiver element comprises an inner PD array and a lens that focuses the light onto the inner array. Multiple receiver elements are combined to create the outer array, enhancing the overall optical gain performance. To maintain the receiver bandwidth, each PD is followed by a transimpedance amplifier (TIA) and various combining techniques can be employed to process the output signals of each PD. The PD side length and the distance between the lens and the array can be optimized subject to specific design constraints. Numerical findings indicate that for a minimum required half-angle FOV of 7.5°, the optimum solution entails a design with a total receiver area of 4 cm^2 and a height of 1.8 mm, resulting in a data rate of 23.82 Gb/s. However, the optimized design involves a significant number of PDs and TIAs, resulting in high hardware and computational complexities. To this end, as an alternative approach, a non-imaging receiver design based on compound parabolic concentrators (CPCs) providing larger optical gain as well as a wider range of acceptance angles is introduced in [18]. In the proposed design, a multi-tier non-imaging ADR architecture based on CPCs is proposed where each CPC is coupled with a PD array. This study shows that an optimized, compact non-imaging ADR using 2×2 PD arrays that has a height of $\leq$0.5 cm and an effective area of $\leq$0.5 cm^2 achieves a data rate of 12 Gb/s with a half-angle FOV of 30° over a 3 m link. Unlike imaging receivers which generally impose strict alignment requirements due to their limited FOV, such a non-imaging receiver offers a simpler design with a lower implementation complexity and ensures a seamless mobility performance within the receiver FOV [17].

3. Networking architecture for laser-based LiFi

In this section, advanced techniques relevant to the architecture of indoor laser-based wireless communication networks are discussed, addressing issues such as cell formation, interference management, resource allocation, and backhauling.

3.1 Cell formation

Using lasers for data transmission results in various features including high data rates, security and energy efficiency. However, the coverage area of the laser is very confined and can be limited to a few centimeters. In this context, laser-based OWC networks can be designed based on the deployment of a large number of APs. Each AP is an array of VCSELs illuminating a confined area referred to as attocell, and a dense laser-based OWC network guarantees seamless transition for mobile users. Considering this cellular network design, inter-cell interference (ICI) among users belonging to different cells must be controlled. In [23, 24], frequency reuse (FR) was proposed to manage the interference among a number of optical cells. Essentially, a set of frequencies is used among adjacent cells, and then, each frequency can be reused by an optical cell located at a farther distance. Therefore, the interference between two optical cells using the same frequency band can be treated as noise. Furthermore, fractional frequency reuse (FFR) was studied in [25, 26], where the coverage area of an optical cell is divided into three regions, applying different frequency reuse factors in each region. Therefore, the users located at the edge do not interfere with each other, while the users located at the center receive minimum interference. In laser-based networks with very small attocells, users with high mobility may be subject to handover every few centimeters, which then results in the need for resilient handover schemes to ensure continuous service. Given this point, a network-centric (NC) approach was proposed in [24, 27, 28] as shown in **Figure 3**, to overcome the limitations mentioned above in terms of the small coverage area of the optical AP and the enhancement of SNR. In the NC design, the whole receiving plane is divided into multiple static cells, each cell with multiple APs that jointly serve multiple users. It is shown that the NC design can minimize the ICI, and therefore, the performance of the optical network can be enhanced considerably compared with cellular networks deploying FR or FFR-based independent transmission per cell. However, the drawbacks of the NC design must be taken into consideration such as the static shape of the NC cells over time as they are formed regardless of the distribution of users, and the NC design might lead to an imbalance in the load among the formed cells where one of the cells might end up serving a high number of users compared to other cells. In contrast, a network topology given by the user centric (UC) perspective was designed in [27–30] to avoid the ICI among multiple optical cells and achieve high data rates. The UC cells in **Figure 3** are characterized by their elastic shapes where they are formed as a function of the distribution of users. Therefore, the UC design is updated over time in response to any changes in the network resulting from the mobility of users. Compared to the NC design, the UC approach involves high complexity and additional signaling. However, it is worth implementing since it enhances the performance of the laser-based OWC network significantly in terms of SNR and data rate.

Figure 4 shows the performances of various optical network formations in an indoor environment designed using MATLAB with dimensions 8 meters $\times$ 8 meters $\times$ 3 meters (Width $\times$ Length $\times$ Height), where 4 $\times$ 4 laser-based APs are deployed on the ceiling to ensure uniform coverage for 20 users distributed on the communication floor. Other parameters such as bandwidth is 1.5 GHz, optical transmit power is calculated as in [13] to ensure eye safety, laser wavelength is 1550 nm, and laser noise is -155 dB/Hz. Considering orthogonal transmission among the users of each optical cell, the UC formation is superior compared to non-unity FR, FFR and NC formations where more than 60% of the users achieve spectral efficiency beyond 15 [bits/s/Hz]. It is worth mentioning that the optimum UC topology has high complexity that increases with the density of the network, while the NC formation is deployed by simply dividing the number of APs into multiple NC cells where each NC cell serves users within its coverage area. However, NC does not eliminate ICI completely, and the complexity of the UC formation can be relaxed by implementing sub optimal solutions such as the well known K-means algorithm.

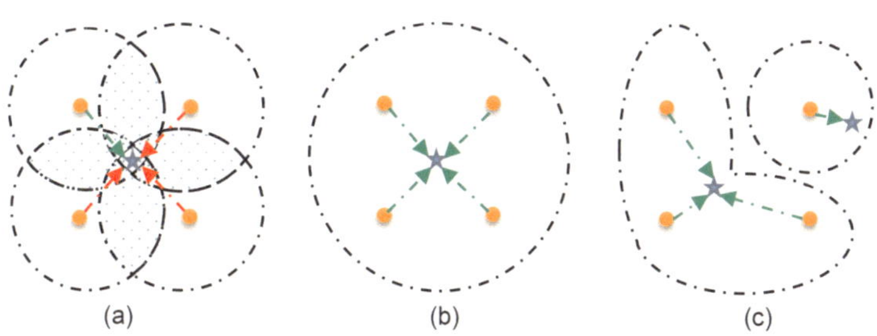

Figure 3.
Cell formation example. (a) Independent transmission per each cell, (b) NC approach, (c) UC approach.

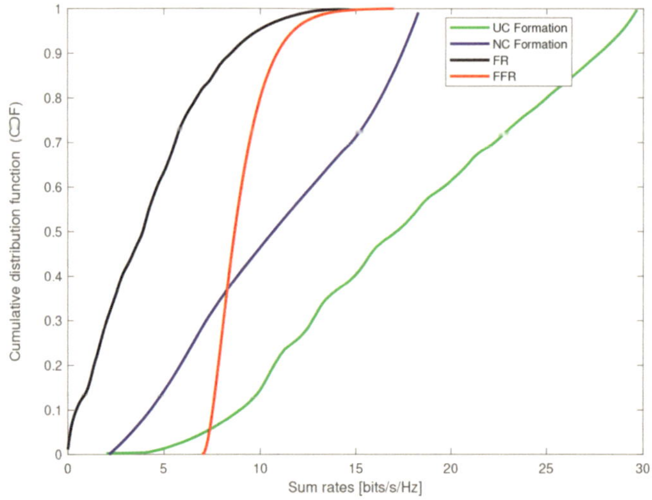

Figure 4.
The CDF curves of the sum rate.

3.2 Interference management

In laser-based OWC networks, serving a high number of users requires efficient interference management schemes to align data streams transmitted to users with a minimum error. Orthogonal transmission schemes such as time division multiple access (TDMA) [31], space-division multiple access (SDMA) [32] and orthogonal frequency division multiple access (OFDMA) [33] were implemented to manage the interference among multiple users through transmitting data streams in an orthogonal fashion giving users the ability to decode their information independently in the absence of interference in their allocated time, frequency or space channels. Despite the low complexity of these schemes, they limit the capacity of laser-based OWC networks due to the fact that users cannot be served over the same frequency or time slots. Another orthogonal approach that is particularly suited for the OWC technology is wavelength division multiplexing (WDM), which was used in [34, 35] by employing RGBY LEDs for data transmission. The WDM can support a number of colors to serve multiple users by allocating a different color to each user. In this context, optical filters must be used on the receiver side to differentiate various wavelengths. However, having many different colors and a broad emission spectrum can result in crosstalk among colors where optical filters cannot distinguish the wavelengths [36]. This issue might not affect the performance of multi-user laser-based OWC networks to the same degree due to the fact that lasers have a narrower emission spectrum. The aim of this subsection is to highlight promising techniques for interference management in laser-based OWC networks.

3.2.1 Precoding schemes

Transmission precoding (TPC) schemes such as minimizing the mean square error (MMSE) [37], zero-forcing (ZF) or interference alignment (IA) [38] derived originally for RF networks were implemented for OWC networks. It is shown that these

TPC schemes can potentially increase the capacity of OWC networks compared to orthogonal transmission schemes. However, they are more suitable for RF networks due to the distinct nature of transmitted RF signals compared to optical signals. For example, TPC schemes might not achieve their full performance of interference management in laser-based OWC networks due to the non-negativity constraint of the optical signal. Given that, TPC schemes must be modified through applying a DC bias current that can eliminate the negative part of the transmitted signal. Therefore, MMSE and ZF schemes were adjusted in [37, 38], respectively, considering the constraints of the optical channel. Furthermore, in [39], a power allocation optimisation problem was formulated in order to maximize the data rate of a TPC-based OWC system constrained by the illumination function. However, encoded information may experience distortion resulting from adding the DC bias current. Moreover, TPC schemes require channel state information (CSI) at the transmitters to maximize the multiplexing gain of the network. Note that providing accurate CSI is not straightforward, and its cost increases considerably with the density of the network. As mentioned, Laser-based OWC networks are expected to be very dense networks, and therefore, the cost and imperfection of CSI must be taken into consideration in such cellular networks. For instance, having a frequency division duplex (FDD) system[1], a fraction of the network resources is dedicated for transmitting estimation pilots. Also, resources must be assigned for feeding back information to the transmitters through the uplink transmission. The estimation pilots are transmitted in an orthogonal fashion over time slots given by the number of transmitters. Therefore, the implementation of TPC schemes in laser-based OWC networks may result in inefficient use of resources.

3.2.2 Non-orthogonal multiple access

A transmission scheme referred to as non-orthogonal multiple access (NOMA) was proposed in [40] to enhance the spectrum efficiency of RF networks. In NOMA, users are grouped based on the channel difference in order to serve them over the same frequency or time slot by allocating different power levels to each data streams. It is shown that NOMA can enhance the capacity of RF systems compared to orthogonal transmission schemes in which exclusive frequency or time resources are assigned to each user. NOMA can be implemented in OWC systems to offer enhanced user experience. However, the fundamental differences between RF and optical systems in terms of the coverage area, number of APs and nature of the transmitted signal result in new challenges. In [41], NOMA was proposed for interference management in an optical system where the sum rate was maximized by tuning the semi-angle of the optical AP and the field of view (FoV) of the receptor. In [42], NOMA operating under illumination constraints shows superior performance compared to other orthogonal transmission schemes. In [43], the performance of NOMA was evaluated in an optical system serving multiple users where each user has a target requirement in terms of QoS.

In a laser-based OWC network, NOMA can be implemented by first sorting the users in ascending order based on their channel gains, and then, forming multiple pairs of strong and weak users using a clustering algorithm. Note that, the interference among the groups must be managed, while NOMA is considered for transmission within each group. The users in NOMA receive power control coefficients based on

[1] The uplink transmission in laser-based OWC networks is usually implemented using frequencies other than optical.

their channel gains. These power coefficients can be determined through optimization problems that aim to maximize the sum rate under fairness between the weak and strong users. In general, NOMA allocates less power to the strong users located close to the transmitters, while more power is allocated to the users with poor channel gains. The strong user must perform successive interference cancelation (SIC) to decode the information with high power first and subtract it from the received signal. This process is followed by the strong user iteratively until it decodes the useful information. On the other hand, the weak user can decode its information from the received signal while managing other information as noise. In extreme scenarios, the strong user can foreword the desired information and server the weak user through other links such as RF networks deployed as complementary networks, i.e., hybrid optical/RF networks.

3.2.3 Blind multiple access

Blind interference alignment (BIA) was proposed for RF networks in [44] in order to align the interference among users without the need for CSI at the transmitters. BIA exploits the channel correlation among users, where each user is equipped with a reconfigurable antenna. In [45], the performance of BIA was compared to ZF taking into consideration the cost of providing CSI at the transmitters. It is shown that BIA is more suitable than ZF in terms of SNR. For OWC systems, a new optical detector referred to as a reconfigurable detector was derived in [46] to implement BIA and avoid the need for CSI at the optical transmitters. The reconfigurable receiver is a set of PDs connected to a single signal processing chain through a selector. Therefore, it has the ability to provide a set of independent channel responses, which are essential for implementing BIA. Interestingly, BIA is characterized by its non-negative precoding matrix given by {0, 1}, which means that the transmitted signal is naturally positive without applying a DC bias current, making BIA more suitable for OWC systems than TPC schemes. However, BIA faces challenges in large scale OWC networks. In particular, the length of the BIA transmission block increases with the number of optical APs and users, and therefore, a large channel coherence time is required to transmit the entire block. In addition, BIA is limited by noise resulting from interference subtraction. In [27, 47], network topologies such as NC and UC designs are implemented to relax the limitations of BIA. It was shown that, BIA based on UC and NC designs achieves higher performance in high density optical networks compared to traditional BIA. **Figure 5** shows the performances of BIA and other interference management schemes in terms of aggregate sum rate in an indoor environment similar to **Figure 4**. It can be seen that BIA achieves a slightly higher sum rate compared to NOMA and TPC schemes. Note that, the application of NOMA requires an advanced power allocation scheme to distribute the power between week and strong users, and the TPC scheme suffers performance degradation in high density networks due to the requirements of accurate CSI. Moreover, BIA results in full connectivity at the cost of increasing noise, which is relaxed in the case of NC-BIA and UC-BIA where BIA coordinates the transmission within each cell rather than the whole network.

3.3 Resource allocation

Given the remarkable developments and rapid evolution of communication networks and computing architectures, resource allocation continues to be a non-trivial matter, due to the fact that users are always demanding better QoS, which is driven by

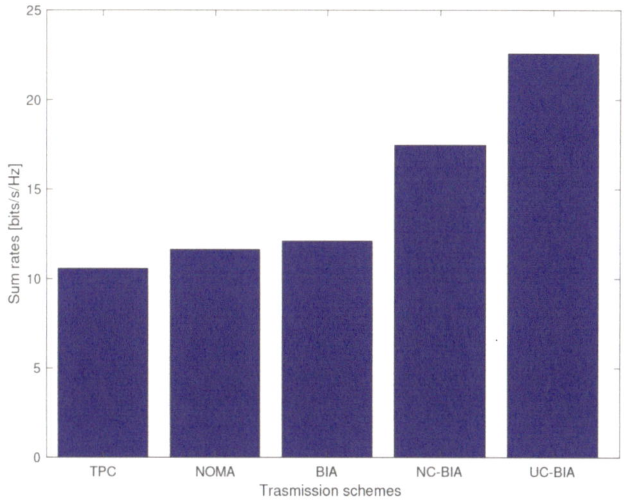

Figure 5.
The sum rates of different interference management schemes.

the increasing demand for services that are computationally intensive and/or bandwidth hungry. The developed resource allocation framework must be able to cope with various new emerging architectures such as wired/wireless mesh networks, multi-protocol networks, cognitive networks, cloud/fog computing systems, etc. Over the recent years, various tools and techniques have been utilized. These tools include optimization theory: centralized solutions such as mixed integer linear programming (MILP) and distributed ones such as game theory, and auction theory, to name a few. Therefore, given the complex nature of communication systems, especially those that comprise of multiple users and heterogeneous set of resources, it becomes imperative to design advanced resource allocation methodologies in order to ensure the optimal performance of such systems whilst users' QoS is guaranteed [48–51]. In this subsection, we focus on an end-to-end cloud/fog infrastructure as shown in **Figure 6**, which consists of multiple OW mobile user devices clustered inside a room forming a mobile fog unit (MobFog). It is assumed that each mobile device can establish a link with one or more light units and is assigned to only one of the available channel wavelengths. All of the APs are connected over a passive optical network (PON) and each AP is connected to an optical network unit (ONU). In return, the OLT aggregates the traffic from all of the ONUs towards other layers in the network. A low-end server, which acts as a fog node (RoomFog) is connected to the OLT device in the same room using a dedicated ONU. Moreover, there are three more fog data centers that are located in the building (BuildingFog), campus (CampFog) and the metro layer (MetroFog). These fog nodes are integrated with the centralized cloud data center (CCloud) over the optical networking infrastructure so that highly demanding requests that cannot be processed by mobile fog or fog nodes, can be processed at the cloud. When a task is allocated to the MobFog for processing, the OLT device sends this task to the participant's mobile devices that are clustered as a fog node via the OWC communication. Then once processing is completed, the output is sent back to the OLT. In other cases where a task has to be assigned to the RoomFog, BuidlingFog, or CampFog nodes, the local Ethernet LAN is used to establish the communication with the required location. However, for processing at the MetroFog or CCloud, demands must be sent over the optical infrastructure to either location.

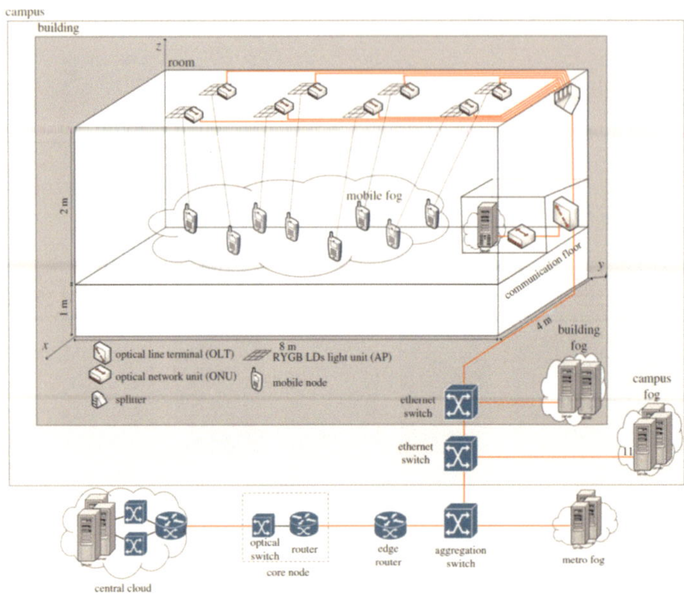

Figure 6.
A cloud/Fog Architecture with OWC Front-End [48].

3.3.1 Cloud/fog network with an OWC front-end

In OWC systems, multiplexing and multiple access techniques are imperative in order to support multiple users. Therefore, to avoid signal quality degradation introduced in a multi-user setting, efficient utilization of resources can mitigate such issues. As for computing, motivated by the goal of reducing latency and power consumption, recently, researchers have focused on emerging distributed architectures with the aim of offloading the processing workloads from the centralized data center. In this direction, distributed "mini" data centers have emerged as the new generation of cloud computing whereby processing demands are "distributed" among fog nodes that are closer to end-user devices. This distributed processing approach results in faster response time in addition to the reduced networking burden. It is important to note that an efficient networking fabric is pivotal for distributed processing. In this regard, OWC systems can offer data rates in a range of Gbps and beyond, thus these mediums can be viewed as promising for supporting the front-end of distributed fog networks in indoor environments [52, 53].

3.3.2 Distributed processing and fog computing

Fog computing is a well-known distributed processing solution that was first coined by Cisco in 2012, and it was aimed at extending the cloud functionalities from the core of the network to the edge of the network, hence closer to the end-user devices [54]. Oftentimes, decision-making can be made better and quicker if the collected data is processed at the edge of the network in close proximity to the source nodes. Currently, fog computing is still in its infancy and a standardized architecture has yet to be agreed. Thus, alternative fog architectures are increasingly being studied in the research community, especially the interplay between the edge devices (fog) and the core (cloud), since fog is regarded as a powerful complement to the cloud. Also, a proper resource allocation scheme is crucial in the fog, as application services can be placed in a highly energy inefficient server or even further from the source node which results in higher communication latency. It is expected that through cooperation between fogs and the centralized cloud, a more efficient and greener computing platform can be achieved [55]. Fog nodes are normally formed by building distributed mini data centers in the network, using a smaller scale of low-end servers [56]. However, due to the unprecedented number of highly intelligent embedded devices at the edge such as IoT devices, which are mostly underutilized has lead to a new definition of fog computing. These IoT devices can collectively provide huge amounts of computational resources when they are clustered together to form a fog mini data center [57].

3.3.3 Optimum resource allocation in OWC networks

Two stages of OWC resource allocation is considered here. In the first stage, the resource allocation problem is modeled in the OWC environment, where the mobile devices are assigned optimally to wavelengths and subsequently to APs. The WDMA or any other advanced technique can be used to support the multiple access of users. It is assumed that the optical line terminal (OLT) inside the room acts as the controller and has global knowledge on users' locations as well as their channel information. The developed mixed integer linear programming (MILP) model in [48] can be used to assign APs and wavelengths to users (end-devices) so that the sum of user SINRs or data rates is maximized. In the second stage, the output of the optimization model in the first stage can be used to jointly optimize the assignment of users to APs and wavelengths and the cloud/fog that involves the optimal placement of processing tasks so that the total power consumption is minimized.

3.4 Backhaul links

Backhaul is an essential component of wireless cellular networks, granting base stations access to the core network. With the continuous trends of cell densification and data rate aggregation and the emergence of real-time bandwidth-hungry services in 5G networks and beyond, devising a cost-effective backhaul solution is a major challenge. Besides, for future 6G wireless systems, indoor Tb/s wireless access systems are envisioned as a means to offload busy data traffics from massively deployed broadband WiFi and mobile cellular networks. This requires an efficient backbone network that can support aggregate data rates of multi-Tb/s from laser-based optical wireless links.

A natural option for realizing high-capacity backhaul links is the distribution of optical fiber cables in the indoor environment where the Tb/s access network is deployed. In a star topology, every AP is individually equipped with an optical fiber link that connects it to the fiber distribution hub through which it is routed to the Internet. However, this is feasible if the expenditure incurred due to redesigning the wiring infrastructure is justifiable. 80% of Internet traffic was reported to initiate and terminate indoors. Thus, it improves such a fiber distributions network if its connections are in the form of a data network (i.e. replacing the star topology with a richer network). Such a network can then meet these indoor demands, which can be in the form of user-to-user and IoT-to-IoT communication in addition to user-to-Internet and vice-versa traffic. **Figure** 7 illustrates how such a network can connect a number of rooms in a building. Improving the connections between APs within a room, floor or a whole building and the final gateway is beneficial for users and service providers as it can reduce both data transmission delay and networking power consumption.

PONs are key technologies for energy efficient solutions. We will also show that among the different options, some solutions are more adaptable than others. Different solutions can be listed as:

- **PON-based on a TDMA technology and adopting a tree topology:** The network is fully passive and minimizes the number of fibers deployed in the field. The drawback is a physical bit rate at the subscriber side that follows the aggregated bit rate. Since the power consumption is sensitive at the periphery of the network, this solution operating at ultra-high bit rates could be limited.

- **Passive Optical networks exploiting WDMA technologies:** Each subscriber has a physical interface corresponding to its guaranteed bandwidth. The aggregation is made with optical multiplexers to transport the data of several subscriber onto a same fiber. This solution can require tunable lasers to adapt the wavelength, in particular, if the transport fiber is shared with another PON system. This solution could be adapted for low power consumption.

- **Fiber to the building solutions:** The transport data is based on TDM techniques. The connections of the trunk may be terminated in the basement of a building or in the street cabinet. The objective here is to disaggregate the bit rate, to offer a physical bit rate at the subscriber side as close as possible to the guaranteed bandwidth. For example, interleaving techniques can be implemented as well as disaggregation techniques.

- **Passive optical LANs (POL):** POLs offer the possibility to distribute the couplers to adapt the connectivity to a room of a space. The main drawback is that the aggregated bit rate is imposed on the subscribed side.

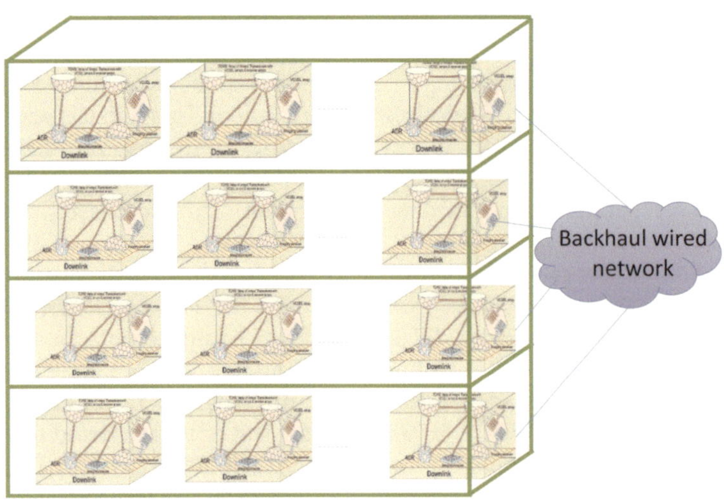

Figure 7.
Wired backhaul network to connect APs in different rooms within a building.

- **Optical ring networks:** They are relevant when there is a risk of fiber cuts. Typically, this technology is better positioned than PONs for outdoor interconnection. It is also a good solution to offer machine-to-machine (M2M) communication (for example in factories). With proper management of the physical layer, it is easier to increase the number of interconnection points, when compared to a PON.

The physical ring topology guarantees the M2M service through the adoption of broadcast and select mechanisms. An optical ring based on optical add/drop multiplexers (OADM) is guaranteed to have a high level of reliability (if a node is in failure the optical ring is not impacted and the traffic is not interrupted). Finally, the ring topology offers natural protection schemes by doubling the ring and by exploiting the bidirectionality. For time sensitive networks, it is proposed to combine OADMs operating in WDM regime, acceleration of the bit rate (from x to $10x$ bit rates), and minimum buffering when possible through native Ethernet switching. The estimation of the end-to-end latency (outside the propagation delay) for a TDMA operation combined with a Broadcast and Select (B&S) mechanism (allowing a multicast or unicast operation) are the followings:

- Insertion delay:
    - One delay to form the slot: $< 1\mu s$ in a B&S mode
    - One delay to insert the slot in the ring: delay $< 10\mu s$
    - total: $< 11\mu s$
- Transit delay:
    - 0 delay, since there is no in line processing for header with the B&S mechanism
- Drop delay:
    - One delay to extract the control channel and synchronize the control channel with the data stream: delay $< 10\mu s$

The total end-to-end latency (without the propagation delay of the ring) is $21\mu s$.

The implementation of multiple topologies can ensure higher sustainability and availability of the network. The use of one topology (tree, bus or ring) cannot be enough to satisfy the exigence of some use cases. It is then mandatory to envisage multiple topologies. For example in factories where the communication of machines is important to control the production process step by step, optical ring networks are required. By adopting broadcast and select techniques it is then possible to offer unicast or multicast connections to enable this machine-to-machine service. If we are in the case of a multiclient factory, then a PON can be important to interconnect the different aggregation points of the ring networks to the PON. This is a typical case where ultra-high bit rate PON will be required. For some other use case like the deployment of OWC AP in a hospital it could be more adapted to deploy a POL per floor and interconnect different POL using a ring network.

As a cost effective solution, the IEEE 802.3bt PoE++ becomes the standard for connected homes and buildings. Proposing a network topology where LiFi access points (APs) connected using a PoE++ Gigabit Ethernet reduces the cost of integration, deployment and maintenance. Keeping the same passive optical LAN topology in **Figure 6**, a new networking option, which consists of an interconnecting optical network unit (ONU) with a PoE++ switch equipped with the small form-factor pluggable (SFP) ports, is suggested by Crantec. This interconnection will then create a 802.3bt PoE++ Gigabit Ethernet LAN backbone to deploy a LiFi wireless network as illustrated in **Figure 8**. In a distributed architecture, LiFi APs can be connected to the 802.3bt PoE++ network using the industrial micro switch. If a VCSEL arrays LiFi AP is then equipped with a PoE network connectivity, it can be easy and at a low cost to connect it to a Gigabit Ethernet switch as shown in the **Figure 9**.

For such a network shown in **Figure 7**, this chapter proposes a WDM-PON network [59], which uses passive Arrayed Waveguide Grating Routers (AWGRs) and an OLT port to realize all-to-all non-blocking connections using a number of predefined wavelengths. An AP can then tune the wavelength of a transceiver to communicate with another AP and directly serve user-to- user traffic. The performance and energy consumption of this proposal was compared to a widely used data network which is a Spine-leaf network design. This network contains two layers of fully connected electronic switches. The comparison was performed under optimized routing for user to user traffic with the objective of minimizing the power consumption or the delay, and WDM-PON networks show superiority over Spine-leaf data networks.

An alternative approach is the use of laser-based optical wireless links to interconnect APs with the core network. It is shown that ultra-high-speed transmission rates of more than 1 Tb/s can be achieved based on VCSEL arrays and multiple-input multiple-output (MIMO) system configurations [60, 61]. Note that, crosstalk in the MIMO channel is one of the challenges in such configurations where the beam spots of

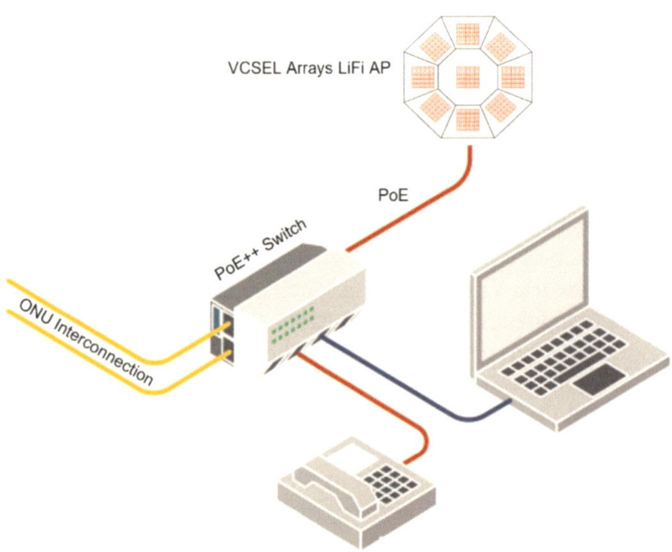

Figure 8.
Indoor PoE++ LiFi wireless LAN network connecting a VCSEL arrays LiFi AP. The LiFi AP shown in this figure is based on an array of arrays of VCSELs referred to as the double-tier AP architecture [58].

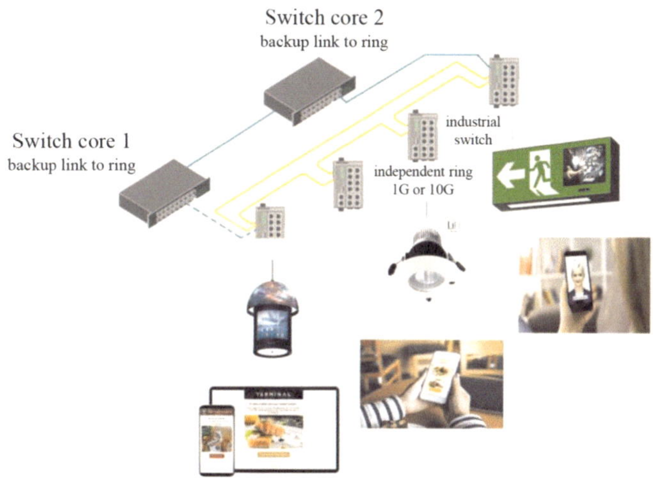

Figure 9.
Indoor PoE++ LiFi wireless LAN network mobile applications & services. The LiFi AP.

VCSELs at the receiver end must be confined. As a consequence, a practical limitation of line-of-sight (LOS) MIMO optical wireless link design with laser beams of very low divergence angles is that it requires precise alignment. Therefore, in realistic scenarios, the system performance is prone to any kind of misalignment arising from installation errors or transceiver vibrations.

4. Conclusions

The chapter provides a thorough study on the use of OWC in 6G wireless networks. We first design new optical transmitters and receivers to support the extreme demands of current and future Internet-based applications. Multiple VCSEL arrays deployed in an indoor environment can ensure uniform coverage in multi-user scenarios and provide aggregate data rate transmission in a range of Tbps. On the receiver end, we introduce an array of PD arrays to enhance the FOV, gain and bandwidth, and to support ultra-high-speed optical links. Finally, various challenges and potential solutions from the networking perspective are discussed to help build an end-to-end OWC infrastructure in real-world deployment of 6G wireless networks.

Acknowledgements

The authors acknowledge financial support from the EPSRC under program grant EP/S016570/1 'Terabit Bidirectional Multi-User Optical Wireless System (TOWS) for 6G LiFi'.

Author details

Ahmad Adnan Qidan[1*], Mohammad Dehghani Soltani[2], Wajahat Ali[3], Rui Chen[3], Shenjie Huang[2], Barzan Yosuf[1], Sanaa Mohamed[1], Ravinder Singh[3,4], Yi Liu[4], Hossein Kazemi[5], Elham Sarbazi[5], Bela Berde[6], Dominique Chiaroni[6], Bastien Bechadergue[7], Fathi Abdeldayem[8], Hardik Soni[9], Jose Tabu[10], Micheline Perrufel[11], Nikola Serafimovski[12], Taisir E.H. El-Gorashi[1], Michael Crisp[3], Richard Penty[3], Ian H. White[3,13], Harald Haas[5], Majid Safari[2] and Jaafar Elmirghani[1]

1 King's Collage London, United Kingdom

2 University of Edinburgh, United Kingdom

3 University of Cambridge, United Kingdom

4 Toshiba Europe Limited, United Kingdom

5 University of Strathclyde, United Kingdom

6 Nokia Bell Labs, United States

7 Oledcomm, France

8 Du AUE, United Arab Emirates

9 Nav Wireless Technologies Pvt Ltd, India

10 Crantec, Switzerland

11 Orange, France

12 pureLiFi Limited, United Kingdom

13 University of Bath, United Kingdom

*Address all correspondence to: ahmad.qidan@kcl.ac.uk

References

[1] Haas H, Yin L, Wang Y, Chen C. What is LiFi? Journal of Lightwave Technology. 2016;**34**(6):1533-1544

[2] Soltani MD, Xiping W, Safari M, Haas H. Bidirectional user throughput maximization based on feedback reduction in LiFi networks. IEEE Transactions on Communications. 2018; **66**(7):3172-3186

[3] Steigerwald DA, Bhat JC, Collins D, Fletcher RM, Holcomb MO, Ludowise MJ, et al. Illumination with solid state lighting technology. IEEE Journal of Selected Topics in Quantum Electronics. 2002;**8**(2):310-320

[4] McKendry JJD, Massoubre D, Zhang S, Rae BR, Green RP, Erdan G, et al. Visible-light communications using a cmos-controlled micro-light-emitting-diode array. Journal of Lightwave Technology. 2012;**30**(1):61-67

[5] Karunatilaka D, Zafar F, Kalavally V, Parthiban R. Led based indoor visible light communications: State of the art. IEEE Communications Surveys & Tutorials. 2015;**17**(3):1649-1678

[6] Xiao H, Wang R, Wang K, Chen W, Chiang KS. Trade-offs between illumination and modulation performances of quantum-dot LED. IEEE Photonics Technology Letters. 2020;**32**(12):726-729

[7] Singh KJ, Fan X, Sadhu AS, Lin C-H, Liou F-J, Tingzhu W, et al. $CsPbBr_3$ perovskite quantum-dot paper exhibiting a highest 3 dB bandwidth and realizing a flexible white-light system for visible-light communication. Photonics Research. 2021;**9**(11): 12002341

[8] O'Brien DC, Faulkner GE, Zyambo EB, Jim K, Edwards DJ, Stavrinou P, et al. Integrated transceivers for optical wireless communications. IEEE Journal of Selected Topics in Quantum Electronics. 2005;**11**(1):173-183

[9] Henderson R, Schulmeister K. Laser Safety. 1st ed. CRC Press; 2003. Available from: https://www.perlego.com/book/1628387/laser-safety-pdf. [Accessed: October 14, 2022]

[10] Wang W-C, Cheng C-H, Wang H-Y, Lin G-R. White-light color conversion with red/green/violet laser diodes and yellow light-emitting diode mixing for 34.8 gbit/s visible lighting communication. Photonics Research. 2020;**8**(8):1398-1408

[11] Zhang L, Wei Z, Wang Z, Geng Z, Wei G, Julian Cheng HY, et al. High-speed multi-user optical wireless communication between vcsel-integrated electronic devices. Optics Communications. 2021;**486**:126774

[12] Wei Z, Guan M, Zang Z, Fu HY. Utilization of 850 nm near-infrared VCSEL for high-capacity indoor free space optical communications. In: Asia Communications and Photonics Conference (ACP), Hangzhou, China. 2018. pp. 1-3. DOI: 10.1109/ACP.2018.8596214

[13] Liu Y, Wajahat A, Chen R, Bamiedakis N, Crisp M, White IH, et al. High-capacity optical wireless vcsel array transmitter with uniform coverage. In: Free-Space Laser Communications XXXV. Vol. 12413. San Francisco, California, United States: SPIE; 2023. pp. 144-150

[14] Soltani MD, Kazemi H, Sarbazi E, El-Gorashi TEH, Elmirghani JMH,

Penty RV, et al. High-speed imaging receiver design for 6g optical wireless communications: A rate-fov trade-off. IEEE Transactions on Communications. 2023;**71**(2):1024-1043

[15] Kahn JM, Barry JR. Wireless infrared communications. Proceedings of the IEEE. 1997;**85**(2):265-298

[16] Sarbazi E, Kazemi H, Soltani MD, Safari M, Haas H. Design Tradeoffs of non-imaging angle diversity receivers for 6G optical wireless access networks. In: 2022 IEEE Global Communications Conference (GLOBECOM). Rio de Janeiro, Brazil; 2022. pp. 419-424

[17] Sarbazi E, Kazemi H, Safari M, Haas H. A robust and compact non imaging angle diversity receiver for 6G optical wireless communications. In: 2023 IEEE International Conference on Communications Workshops (ICC Workshops). Rome, Italy; 2023. pp. 1486-1491

[18] Sarbazi E, Kazemi H, Crisp M, El Gorashi T, Elmirghani J, Penty R, et al. Design and optimisation of high-speed receivers for 6G optical wireless networks. IEEE Transactions on Communications (Early Access). DOI: 10.1109/TCOMM.2023.332865

[19] Sarbazi E, Kazemi H, Soltani MD, Safari M, Haas H. Imaging angle diversity receiver design for 6G optical wireless communications: Performance Tradeoffs and optimisation. In: 2023 IEEE Global Communications Conference (GLOBECOM). Kuala Lumpur, Malaysia; 2023

[20] Alkhazragi O, Trichili A, Ashry I, Ng TK, Alouini M-S, Ooi BS. Wide-field-of-view optical detectors using fused fiber-optic tapers. Optics Letters. 2021;**46**(8):1916-1919

[21] Koonen T, Mekonnen K, Huijskens F, Cao Z, Tangdiongga E. Novel broadband owc receiver with large aperture and wide field-of-view. In: 46th European Conference on Optical Communications (ECOC 2020), Brussels, Belgium. 2020. pp. 1-4

[22] Umezawa T, Matsumoto A, Akahane K, Nakajima S, Yamamoto N. Large submillimeter high-speed photodetector for large aperture fso receiver. IEEE Journal of Selected Topics in Quantum Electronics. 2021;**28**(2: Optical Detectors):1-9

[23] Kim H, Kim D, Yang S, Son Y, Han S. Mitigation of inter-cell interference utilizing carrier allocation in visible light communication system. IEEE Communications Letters. 2012;**16**(4): 526-529

[24] Li X, Zhang R, Hanzo L. Cooperative load balancing in hybrid visible light communications and wifi. IEEE Transactions on Communications. 2015; **63**(4):1319-1329

[25] Chen C, Serafimovski N, Haas H. Fractional frequency reuse in optical wireless cellular networks. In: 2013 IEEE 24th Annual International Symposium on Personal, Indoor, and Mobile Radio Communications (PIMRC). London, UK; 2013. pp. 3594-3598

[26] Chen C, Videv S, Tsonev D, Haas H. Fractional frequency reuse in dco-ofdm-based optical attocell networks. Journal of Lightwave Technology. 2015;**33**(19): 3986-4000

[27] Adnan-Qidan A, Morales Céspedes M, García Armada A. User-centric blind interference alignment design for visible light communications. IEEE Access. 2019;**7**:21220-21234

[28] Adnan-Qidan A, Morales-Céspedes M, Armada AG. Load

balancing in hybrid vlc and rf networks based on blind interference alignment. IEEE Access. 2020;**8**:72512-72527

[29] Zhang R, Wang J, Wang Z, Xu Z, Zhao C, Hanzo L. Visible light communications in heterogeneous networks: Paving the way for user-centric design. IEEE Wireless Communications. 2015;**22**(2):8-16

[30] Li X, Jin F, Zhang R, Wang J, Xu Z, Hanzo L. Users first: User-centric cluster formation for interference-mitigation in visible-light networks. IEEE Transactions on Wireless Communications. 2016;**15**(1): 39-53

[31] Abdelhady AM, Amin O, Chaaban A, Shihada B, Alouini M. Downlink resource allocation for dynamic tdma-based vlc systems. IEEE Transactions on Wireless Communications. 2019;**18**(1): 108-120

[32] Chen Z, Haas H. Space division multiple access in visible light communications. In: 2015 IEEE International Conference on Communications (ICC). London, UK; 2015. pp. 5115-5119

[33] Lin B, Tang X, Ghassemlooy Z, Lin C, Li Y. Experimental demonstration of an indoor vlc positioning system based on ofdma. IEEE Photonics Journal. 2017;**9**(2):1-9

[34] Cossu G, Khalid AM, Choudhury P, Corsini R, Ciaramella E. 2.1 gbit/s visible optical wireless transmission. In: 2012 38th European Conference and Exhibition on Optical Communications. Amsterdam, Netherlands; 2012. pp. 1-3

[35] Wang Y, Tao L, Huang X, Shi J, Chi N. 8-gb/s rgby led-based wdm vlc system employing high-order cap modulation and hybrid post equalizer. IEEE Photonics Journal. 2015;**7**(6):1-7

[36] Singh R, O'Farrell T, David JPR. Analysis of forward error correction schemes for colour shift keying modulation. In: 2015 IEEE 26th Annual International Symposium on Personal, Indoor, and Mobile Radio Communications (PIMRC). Hong Kong, China; 2015. pp. 575-579

[37] Li B, Wang J, Zhang R, Shen H, Zhao C, Hanzo L. Multiuser miso transceiver design for indoor downlink visible light communication under per-led optical power constraints. IEEE Photonics Journal. 2015;**7**(4):1-15

[38] Shen H, Deng Y, Xu W, Zhao C. Rate-maximized zero-forcing beamforming for vlc multiuser miso downlinks. IEEE Photonics Journal. 2016;**8**(1):1-13

[39] Jiang R, Wang Z, Wang Q, Dai L. Multi-user sum-rate optimization for visible light communications with lighting constraints. Journal of Lightwave Technology. 2016;**34**(16): 3943-3952

[40] Saito Y, Kishiyama Y, Benjebbour A, Nakamura T, Li A, Higuchi K. Non-orthogonal multiple access (Noma) for cellular future radio access. In: 2013 IEEE 77th Vehicular Technology Conference (VTC Spring). Dresden, Germany; 2013. pp. 1-5

[41] Marshoud H, Kapinas VM, Karagiannidis GK, Muhaidat S. Non-orthogonal multiple access for visible light communications. IEEE Photonics Technology Letters. 2016;**28**(1):51-54

[42] Kizilirmak RC, Rowell CR, Uysal M. Non-orthogonal multiple access (Noma) for indoor visible light communications.

In: 2015 4th International Workshop on Optical Wireless Communications (IWOW). Istanbul, Turkey; 2015. pp. 98-101

[43] Yin L, Popoola WO, Wu X, Haas H. Performance evaluation of non-orthogonal multiple access in visible light communication. IEEE Transactions on Communications. 2016;**64**(12): 5162-5175

[44] Gou T, Wang C, Jafar SA. Aiming perfectly in the dark-blind interference alignment through staggered antenna switching. IEEE Transactions on Signal Processing. 2011;**59**(6):2734-2744

[45] Wang C, Papadopoulos HC, Ramprashad SA, Caire G. Improved blind interference alignment in a cellular environment using power allocation and cell-based clusters. In: 2011 IEEE International Conference on Communications (ICC). Kyoto, Japan; 2011. pp. 1-6

[46] Morales-Céspedes M, Paredes-Paredes MC, García Armada A, Vandendorpe L. Aligning the light without channel state information for visible light communications. IEEE Journal on Selected Areas in Communications. 2018;**36**(1):91-105

[47] Qidan AA, Morales-Céspedes M, Armada AG, Elmirghani JMH. User centric cell formation for blind interference alignment in optical wireless networks. In: ICC 2021 - IEEE International Conference on Communications. Montreal, QC, Canada; 2021. pp. 1-7

[48] Alsulami OZ, Alahmadi AA, Saeed MSH, Sarah OM, El-Gorashi TEH, Alresheedi MT, et al. Optimum resource allocation in optical wireless systems with energy-efficient fog and cloud architectures. Philosophical Transactions of the Royal Society A: Mathematical, Physical and Engineering Sciences. 2020; **378**(2169):20190188

[49] Qidan AA, El-Gorashi T, Elmirghani JMH. Artificial neural network for resource allocation in laser based optical wireless networks. In: ICC 2022 - IEEE International Conference on Communications. Seoul, Republic of Korea; 2022. pp. 3009-3015

[50] Qidan AA, Morales-Céspedes M, El Gorashi T, Elmirghani JMH. Resource allocation in laser-based optical wireless cellular networks. In: 2021 IEEE Global Communications Conference (GLOBECOM). Madrid, Spain; 2021. pp. 1-6

[51] Qidan AA, Morales-Céspedes M, Armada AG, Elmirghani JMH. Resource allocation in user-centric optical wireless cellular networks based on blind interference alignment. Journal of Lightwave Technology. 2021;**39**(21): 6695-6711

[52] Hussein AT, Alresheedi MT, Elmirghani JMH. 20 gb/s mobile indoor visible light communication system employing beam steering and computer generated holograms. Journal of Lightwave Technology. 2015;**33**(24): 5242-5260

[53] Hussein AT, Alresheedi MT, Elmirghani JMH. 25 gbps mobile visible light communication system employing fast adaptation techniques. In: 2016 18th International Conference on Transparent Optical Networks (ICTON). Trento, Italy; 2016. pp. 1-7

[54] Mouradian C, Naboulsi D, Yangui S, Glitho RH, Morrow MJ, Polakos PA. A comprehensive survey on fog computing: State-of-the-art and research challenges. IEEE Communications Surveys Tutorials. 2018;**20**(1):416-464

[55] Pengfei H, Dhelim S, Ning H, Qiu T. Survey on fog computing: Architecture, key technologies, applications and open issues. Journal of Network and Computer Applications. 2017;**98**:27-42

[56] Witkowski M, Brenner P, Jansen R, Go DB, Ward E. Enabling sustainable clouds via environmentally opportunistic computing. In: 2010 IEEE Second International Conference on Cloud Computing Technology and Science. Indianapolis, IN, USA; 2010. pp. 587-592

[57] Kuada E, Olesen H. Incentive mechanisms for opportunistic cloud computing services. In: 8th International Conference on Collaborative Computing: Networking, Applications and Worksharing (CollaborateCom), Pittsburgh, PA, USA; 2012. pp. 127-136

[58] Sarbazi E, Kazemi H, Soltani MD, Safari M, Haas H. A Tb/s indoor optical wireless access system using VCSEL arrays. In: 2020 IEEE 31st Annual International Symposium on Personal, Indoor and Mobile Radio Communications. London, UK; 2020. pp. 1-6

[59] Elmirghani J, El-Gorashi T, Hammadi A. Passive Optical-Based Data Center Networks. WO Patent App; 2016. p. PCT/GB2015/053,604

[60] Kazemi H, Sarbazi E, Soltani MD, Safari M, Haas H. A tb/s indoor optical wireless backhaul system using vcsel arrays. In: 2020 IEEE 31st Annual International Symposium on Personal, Indoor and Mobile Radio Communications. London, UK: IEEE; 2020. pp. 1-6

[61] Kazemi H, Sarbazi E, Soltani MD, El-Gorashi TEH, Elmirghani JMH, Penty RV, et al. A Tb/s indoor MIMO optical wireless backhaul system using VCSEL arrays. IEEE Transactions on Communications. 2022;**70**(6):3995-4012

Chapter 3

Free Space Optics

Mazen Abdullatif

Abstract

When we design an optical system, we need to consider that the transmitting power is appropriate to reach the receiving end, if this power is less than required, it will be difficult to separate the data from the noise, which in turn, will causes errors. In this chapter, a new method is presented to improve the received power in wireless optical communications by introducing an adaptive power method that adjusts the amplifier's power according to weather conditions. The simulation was done using OptiSystem programming environment and MATLAB environment. Results have shown improvement in the received power, bit error rate, quality factor, and eye diagram, in which the quality factor was improved by a rate ranging from 3.6 to 44.45% in different weather conditions.

Keywords: free space optics (FSO), wireless optical communications, optical amplifiers, EDFA, eye diagram

1. Introduction

Free Space Optics (FSO) technology is very promising in the world of optical communications where the signal is transmitted from the transmitter to the receiving end through free space and using the laser beam.

This technology has many advantages such as high data rates, faster installation, unlicensed spectrum data, high security, and low cost [1–3].

The most prominent applications of FSO are satellite communications, data center networks, underwater communications, and mobile backhaul (**Figure 1**) [4–7].

The main challenge of FSO system is the different weather conditions that affect the optical link. Attenuation is caused by weather changes such as rain, snow, fog, and dust, which in turn, affects the received power, that is reflected in BER and quality factor [8].

2. Photo transmitters

The light sources convert the electrical signal into a light signal with the provision of sufficient power to overcome the damping to reach the receiving end with a level of power higher than the sensitivity of the receiver. The light source is chosen according to the communication system to be established, that is according to the transmission rate and the distance to be covered. The adjustment is made directly or indirectly (external rate).

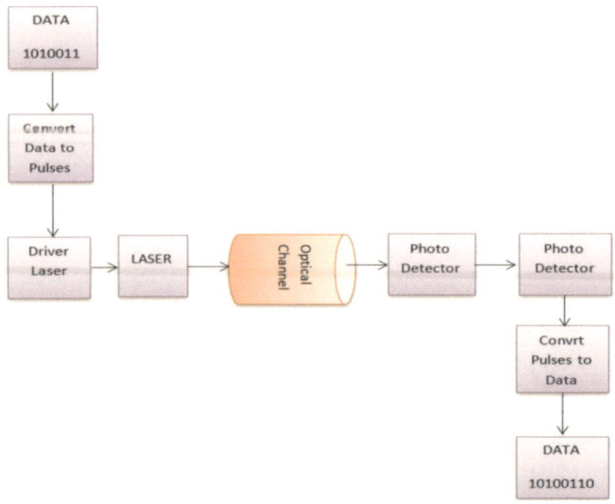

Figure 1.
Wireless optical communications.

Semiconductor laser sources have been adopted in modern optical communications because of their many advantages compared to other laser transmitters, including [9]:

- Small size: These transmitters are built on a single chip that includes all the components needed to emit laser light, which allows them to be easily combined with other equipment.
- Great efficiency: The efficiency of the laser transmitter is up to 50%, which allows it to be driven with low electrical power compared to other transmitters.
- It is driven directly and with a small current: that is, conventional transistor circuits can be used to power the laser transmitter.
- High reliability.

However, these transmitters are not without some disadvantages, including high sensitivity to heat, and therefore the issue of cooling is a fundamental issue that must be taken into consideration.

2.1 Basic parameters of laser transmitters

- Transmitter output power: It is measured in mW or dBm.
- The wavelength of light emitted by the transmitter, as the laser transmission takes place at a wavelength that represents one of the transmitter windows.
- Spectral width of the source: it is the width of the output pulse and is usually denoted by $\Delta\lambda$.
- Primary page.

3. Optoelectronic modifiers

3.1 Photoelectric effect

The photoelectric effect occurs when an electric field is applied to a transparent medium, where the optical properties of this medium (the refractive index) change according to the changes of the field applied to the transparent medium [10].

The applied electric field causes changes in the lengths and directions of the axes of the ellipse by the crystal's refractive index. This leads to a change in the polarization of the light wave passing through the crystal. Thus, the optical intensity at the output will change according to the applied voltage, and this is known as the photoelectric effect.

3.2 Photo modulation mechanism

The optical intensity of the laser transmitter can be modified in one of two ways [11]:

3.2.1 A direct modification

This is done by feeding the laser transmitter directly with an electrical information signal. Despite the simplicity of this modification, it is rarely used because of the frequency chirping resulting from the direct modification of the source.

3.2.2 Indirect modification (external)

The semiconductor laser is fed with a continuous electric current to generate a continuous light wave, and the optical power is modified by an external modifier, such as the Mach Zehnder rate, and **Figure 2** shows the external modulation.

3.2.2.1 Mach-Zehnder modulator

This rectifier consists of a planar waveguide on a substrate with a pair of electrodes, to which modulating electrical forces are applied. The instantaneous electric

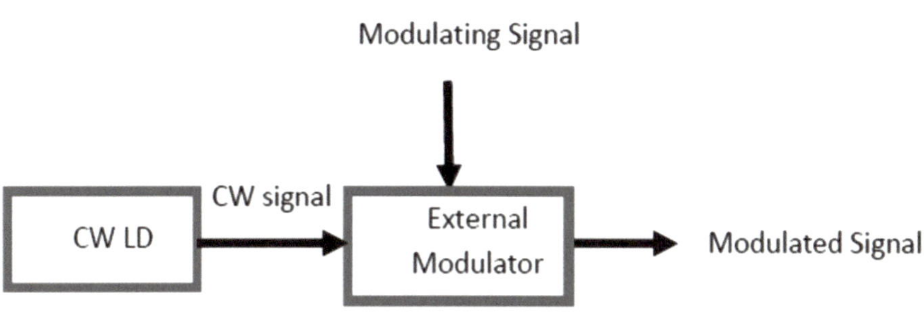

Figure 2.
External modification.

field at the output Eout(t) is given in terms of the instantaneous electric field at the input Ein(t) in relation (1) [11]:

$$Eout(t) = \frac{1}{2}\left[\exp\left(j\frac{\pi}{V_\pi}V1(t)\right) + \exp\left(j\frac{\pi}{V_\pi}V2(t)\right)\right]E_{in}(t) \quad (1)$$

$$V_\pi = V_1 + V_2 \quad (2)$$

In order to obtain a direct relationship between the two optical intensities at the input and output, voltages V1 and V2 are chosen as follows:

$$V1(t) = V(t) - \frac{V_\pi}{2} \quad (3)$$

$$V2(t) = -V(t) + \frac{V_\pi}{2} \quad (4)$$

Thus, the ratio of the electric fields at the input and output is:

$$\frac{Eout(t)}{Ein(t)} = \frac{\sin(\pi.V(t))}{V_\pi} \quad (5)$$

Thus, the optical intensity (the square of the electric field) is modified according to the voltages applied to the two rectifier paths. **Figure 3** shows the rate structure and the main intervention:

3.2.3 Photoreceptors

The photoreceptor converts the optical signal into an electrical signal. There are several types of these receivers depending on their physical structure, but the most common are the photodiode (PIN) and the diode scattering (APD). These elements are characterized by their small size, good sensitivity, fast response, and low noise.

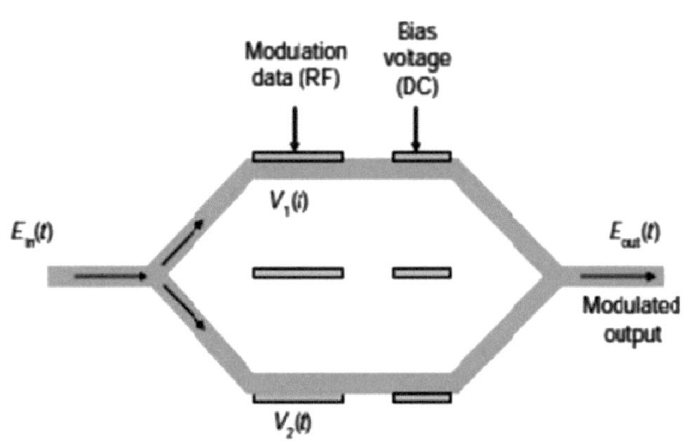

Figure 3.
Mach-Zehnder Modulator.

3.3 Basic parameters of photoreceptors

The photodetector has the following properties:

- The sensitivity of the photodetector: It is defined as the ratio of the output current of the detector (estimated in mA) to the power of the input light (estimated in mW).
- Dark current: One of the specifications of photoreceptors is that they allow the passage of electric current even in the absence of the optical signal (darkness), which leads to the appearance of a dark current that enters as a systematic error in any of the measurements taken by this detector. The cause of this current is the thermal effect generated charge carriers.
- Responsiveness: It is the ratio of the detector's output current to the input light power, measured in A/Watt, and given for a specific wavelength.
- Response speed: It determines the maximum possible transmission speed for a particular photodetector.
- Spectral response: It determines the relationship of the detector's response with wavelength.

4. Photodetector

The primary function of the photodetector is to convert the light signal falling on it into an electrical signal from which the transmitted information can be extracted using suitable electronic circuits. Semiconductor photodetectors depend in their work on the phenomenon of semiconductor absorption of photons, which leads to the generation of free charge carriers that cause the passage of an electric current, and this phenomenon is called the photoelectric effect. If a reverse bias is applied to the P-N junction and without light, a small current will pass through it called the dark current. On the other hand, when the light ray falls on the diode, it absorbs the energy of a photon and an electron-hole pair is formed. If such carriers are formed in the depletion region or near it, it will travel across the junction as a result of the effect of the electric field, and this movement of charge carriers across the junction will lead to the flow of a current in the external circuit of the diode that is directly proportional to the capacity of the light absorbed by the diode, and thus the light is converted into an electric current.

It is a mechanism that gives us analysis of high-speed digital signals. This diagram is formed from the digital form of the signal by folding the part of the signal corresponding to each bit on one graphic curve so that its horizontal axis is time and its vertical axis is the amplitude of the signal, and by repeating this process for many wave samples, and the resulting curve will present the statistical average of the signal, which is similar in shape to the shape of the eye, where **Figures 1–12** shows the ideal shape of the eye chart and the practical form of the eye chart.

5. Eye chart basics

Figure 5 shows the measurements that can be obtained from an eye chart.

One level:

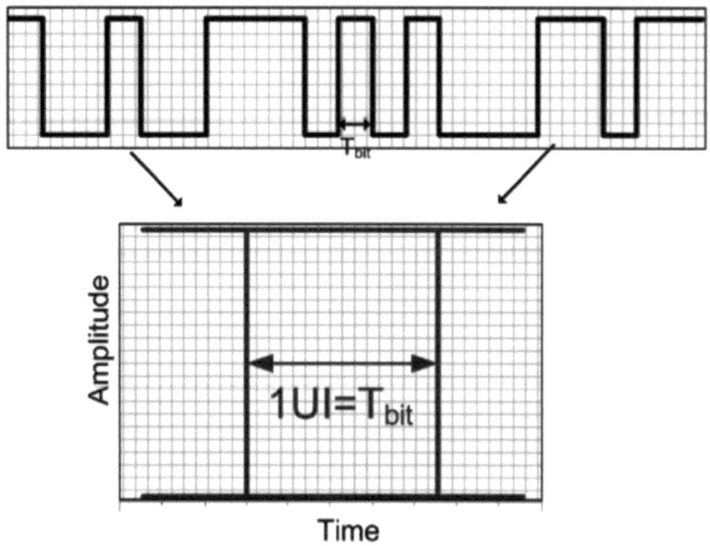

Figure 4.
A typical high-speed digital signal with an eye diagram.

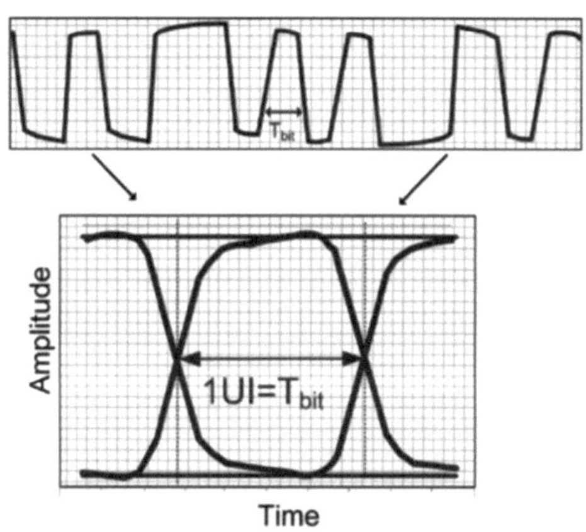

Figure 5.
Practical high-speed digital signal with eye diagram.

The one level in the eye chart expresses the mean value of the one logical amplitude.

Zero level:

The zero level in the eye chart expresses the mean value of the zero logical amplitude.

Eye amplitude:

It is the difference between the one level and the zero level, and based on the amplitude of the eye, the logical circuits at the receiving end will determine if the receiving bit is 1 or 0.

Eye height:

It is a measure of the vertical aperture of the eye chart, the eye height measurement will be equal to the eye amplitude measurement, but for a true eye chart noise will affect the eye height and cause the eye to close, as a result, the eye height

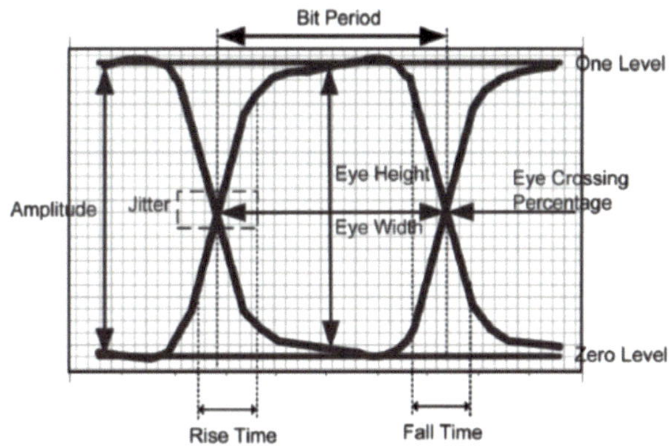

Figure 6.
Scheme of the eye with the measurements that we can read on it.

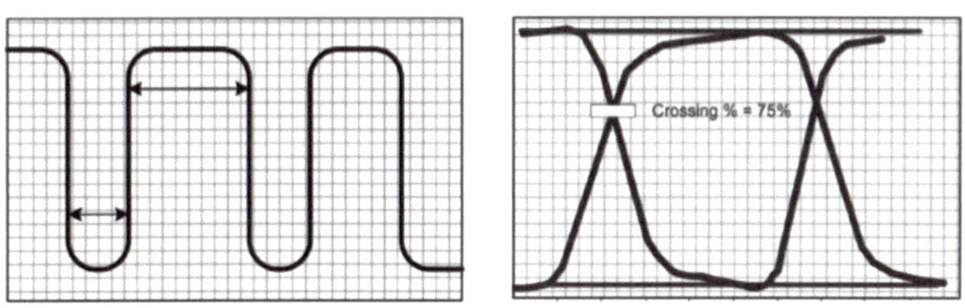

Figure 7.
Format of a received digital signal and the resulting eye diagram with a crossover ratio of 75.

measurement determines the degree of eye closure because of the noise, the SNR indicates directly the degree of eye closure.

Eye crossing percentage:

The eye crossing ratio gives an indication of the distortion of the duty cycle or problems in the symmetry of the pulses at high speeds. The best value for eye crossing percentage is 50% where the duration of the positive pulse is equal to the duration of the negative pulse.

Figure 7 shows the distortion in the duty cycle and its reflection on the eye chart.

6. Effect of weather factors on FSO link

One of the challenges facing the FSO channel is atmospheric attenuation due to absorption, scattering, and scintillation (which may cause signal loss and optical link failure occur), and the effect of these factors changes with time and depends on local conditions and distance. Water molecules and carbon dioxide mainly cause absorption of light signals, while fog, rain, snow, and cloud cause scattering of light signals traveling in free space. This scattering leads to the deviation the light beam that is transmitted from the sender to the receiver. Atmospheric attenuation is given by Beer law, which is the ratio between the power of the transmitted signal and the power of the received signal [8, 12]:

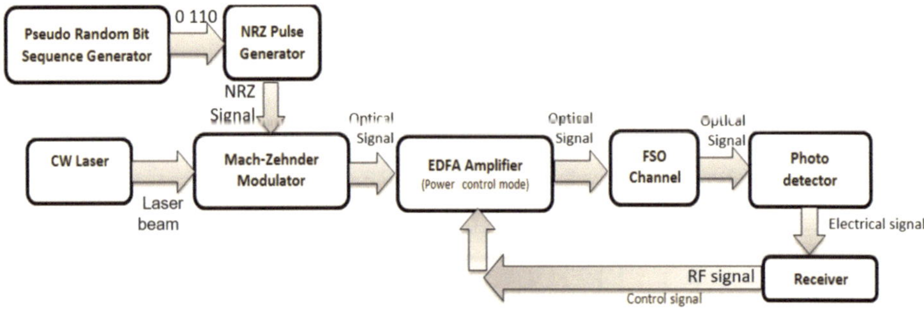

Figure 8.
Proposed method.

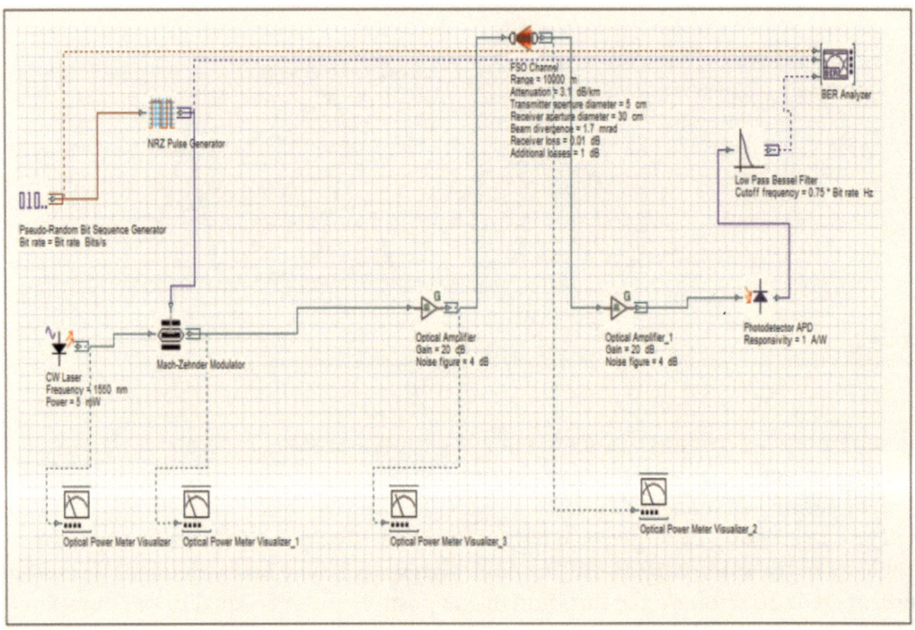

Figure 9.
The environment of simulation used.

$$\tau = P_L = P_0 . \exp(-\alpha . L) \tag{6}$$

Where
P_L: is laser power at the distance L.
P_0: is laser power at source.
α: is air attenuation coefficient.
Where the air attenuation coefficient is calculated by the Kim model.

$$\alpha = \frac{3.91}{V} . \left(\frac{550\text{nm}}{\lambda} \right)^{-q} \tag{7}$$

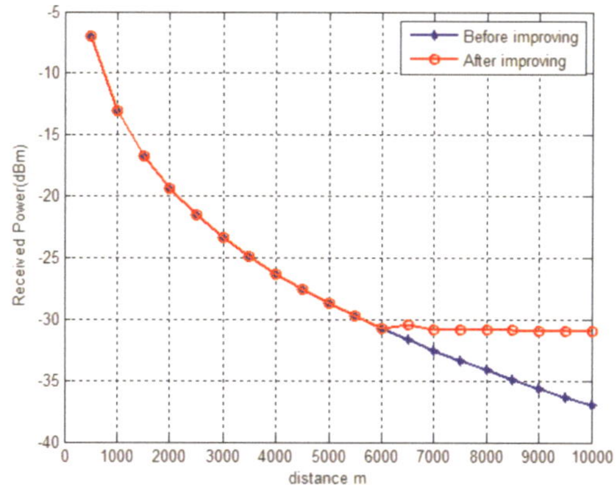

Figure 10.
Received power before and after applying the proposed method for 0.47 dB/km attenuation.

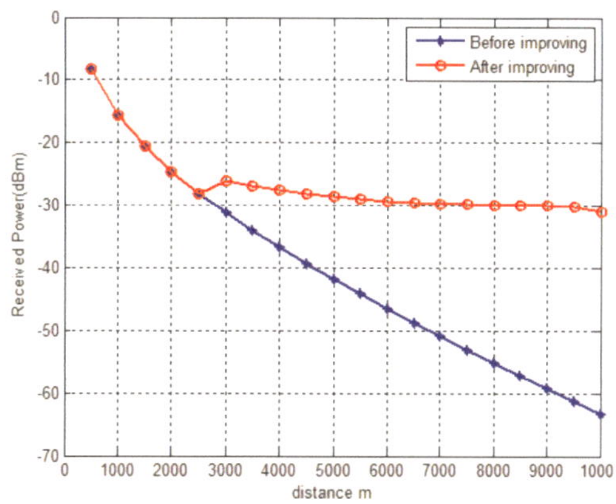

Figure 11.
Received power before and after applying the proposed damping method of 3.1 dB/km.

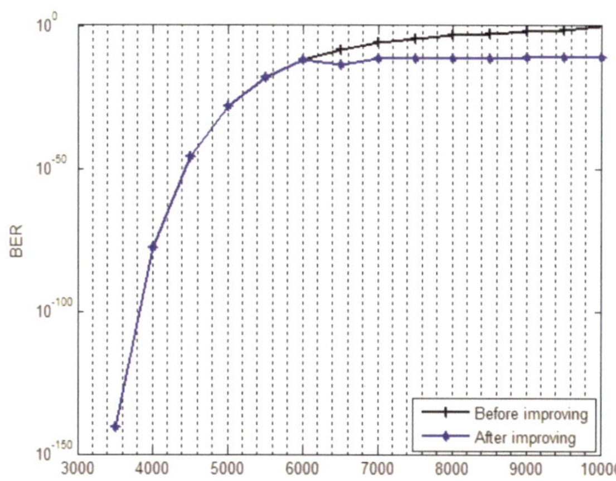

Figure 12.
BER before and after applying the proposed method for damping 0.47 dB/km.

Where

λ: is wavelength in nm.

V:is visibility.

q: is size distribution of the scattering particles.

The attenuation of the transmitted signal can be obtained from the previous model for different weather conditions using q which takes the following values depending on visibility [4, 5]:

q = 1.6 for visibility (V > 50 km).

q = 1.3 for visibility (6 km < V < 50 km).

q = 0.16 V + 0.34 for visibility(1 km < V < 6 km).

q = V − 0.5 for visibility (0.5 km < V < 1 km).

q = 0 for visibility (V < 0.5 km).

7. Optical amplifiers

The importance of optical amplifiers is that they restore optical power without any conversion electro-optical, according to which the principle of the optical amplifier works depends on the induced emission mechanism. There are many types of optical amplifiers, the most important of which are semiconductor amplifiers and similar fiber amplifiers. EDFA-like fiber amplifiers are of great importance in contemporary applications due to their important properties, and they have many work patterns that we will use in our research. This power control mode considers the value of the output power to control the amplifier performance (Power Control). The specified amplifier output power (P_{speci}) is [13–15]:

$$P_{speci} = G \times \sum_{\lambda} P_{in}(\lambda) - \int_{-\infty}^{+\infty} S_{ASE}(f)df \tag{8}$$

where

G: the amplifier gain.

$S_{ASE}(f)$: the spectral density of amplified spontaneous emission integrated on the optical frequency f.

8. The proposed FSO system

When designing an optical system, it must be taken into account that the transmitting power is sufficient to reach the receiving end, and in the event that the receiving power is less than the required limit, it will be difficult to separate the data from the noise, and therefore this matter will cause errors, and to design a well-functioning optical system, considerations must be taken following:

- The received power must be large enough to keep the BER at a low value.
- The receiving power must be small enough to avoid damage to the receiver.
- The transmitted power must be the lowest value required for the system to avoid excessive power consumption.

It was found through studies that when the received power decreases, the bit error rate becomes bad and thus this problem is solved either by increasing the transmitter power or by using an optical amplifier, however if the power increases by a fixed amount, we will have two additional problems, first one is the excessive consumption of power since the weather factors to which the light link is exposed are temporary factors that continue for specific hours of the day, and during certain times of the year. The second is the damages the receiver as a result of permanent exposure to high power.

Hence the idea of this chapter is to suggest an adaptive method of transmitting power according to the weather conditions to which the optical link is exposed. The proposed method ensures a balance between the receiving power being large enough to maintain the BER at a low value in addition to the receiving power being small enough to avoid damaging the receiver.

If the received signal is less than -30 dBm, it sends a control signal from the receiver to the EDFA amplifier which adjusts its power to obtain a good signal reception as shown in **Figure 8**.

9. Results and discussion

The simulation was done using OptiSystem and MATLAB.

OptiSystem is a powerful software for experimenting and controlling fiber optic networks, and optical wireless communication, and has good features for this work.

In this section, numerical results of performance of FSO link based on the proposed system model above according to the parameters shown in **Table 1** are presented for different attenuation and distances.

Performance was evaluated by the received power, bit error rate, and quality factor as shown in Eqs. (9)-(12) [16, 17].

$$P_R = P_T \frac{d2^2}{(d1 + D.R)^2} 10^{(\gamma.R/10)} \qquad (9)$$

Parameter	Typical value
Wavelength	1550 nm
Beam divergences	1.7 mrad
Transmission power	5 mW
Transmitter and receiver loss	0.01 dB
Attenuation	(3.1–0.47)dB/km
Gain amplifier	40 dB
Dynamic range	Up to10000 m
Transmitter hole diameter	5 cm
Receiver hole diameter	30 cm
Optical amplifier noise number	4 dB

Table 1.
Simulated parameters used in the designed optical link.

Where
P_R: is received power.
P_T : is transmitted power.
d1: transmit aperture diameter.
d2: receive aperture diameter.
D: beam divergences.
R: range.
τ: atmospheric attenuation factor.

$$BER = \frac{\mathrm{erfc}\left(\frac{Q}{\sqrt{2}}\right)}{2} \quad (10)$$

Where
BER: is bit error rate.
Q : is quality Factor.
Erfc: is Gauss error function.

$$Q = \frac{SNR\sqrt{2TB}}{1+\sqrt{1+2SNR}} \quad (11)$$

Where
TB: is the bit period.
SNR: signal-to-noise ratio.

$$\mathrm{SNR} = 10\log_{10}\frac{P_s}{P_n} \quad (12)$$

Ps: is signal power.
Pn: is signal noise.

The results show a clear improvement in the received power values, bit error rate, quality factor, and eye diagram. For attenuation value 0.47 dB / km the received power value improved to −30.424 dBm at the distance 6500 m (compared to the

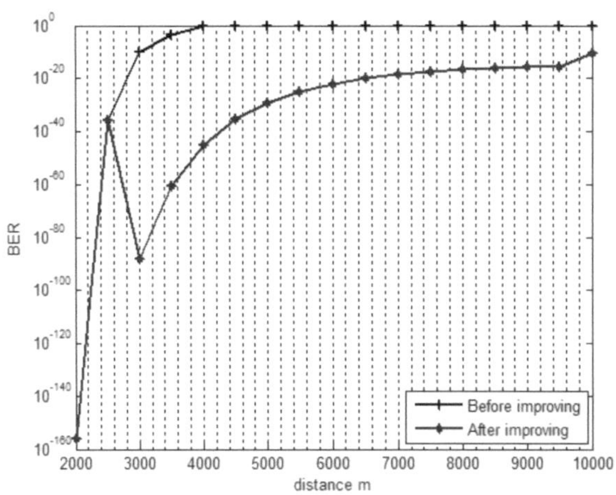

Figure 13.
BER before and after applying the proposed method for damping 3.1 dB/km.

value −31.658 dBm before improvement). At the distance 10,000 m the value of the received power improved to −30.945 dBm (compared to the value −37.031 dBm before improvement) as shown in **Figure 10**.

For attenuation value 3.1 dB/km the power value improved to −25.997 dBm at the distance 3000 m (compared to the value −31.232 dBm before improvement). At the distance 10,000 m the value of the received power improved to −31.094 dBm (compared to the value −63.331 dBm before improvement as shown in **Figure 11**).

For attenuation value 0.47 dB / km the BER value improved to 1.71825e-014 at the distance 6500 m (compared to the value 3.63577e-009 before improvement). At the distance 10,000 m the value of the BER improved to 7.0694e-012 (compared to the value (1) before improvement) as shown in **Figure 12**.

For attenuation value 3.1 dB/km the BER value improved to 6.15722e-089 at the distance 3000 m (compared to the value 1.0669e-010 before improvement). At the

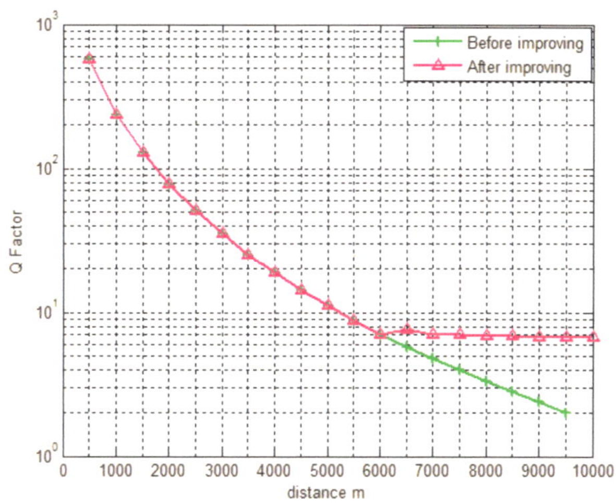

Figure 14.
Quality factor before and after applying the proposed method for damping 0.47 dB/km.

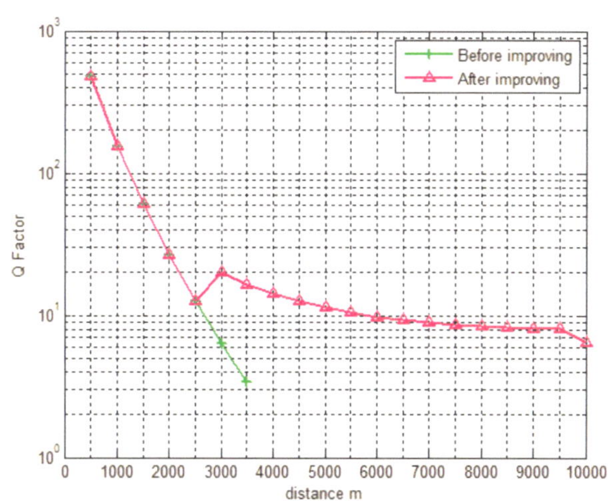

Figure 15.
Quality factor before and after applying the proposed method for damping 3.1 dB/km.

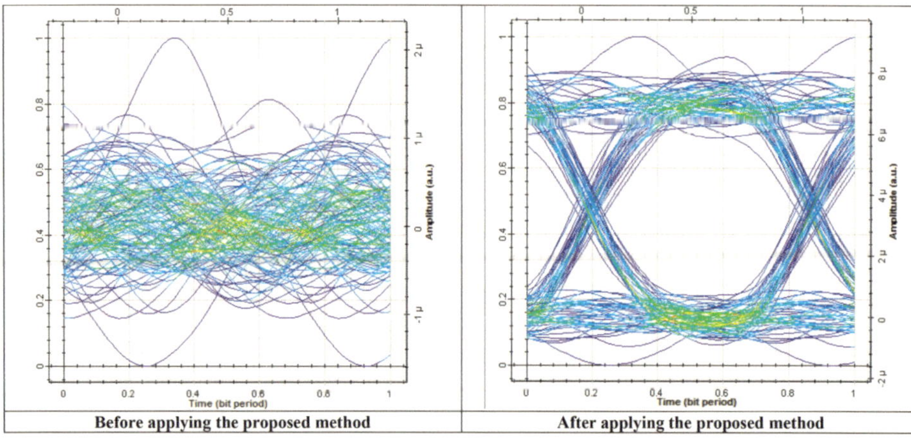

Figure 16.
Eye diagram before and after the application of the proposed method at damping 3.1 and the distance 6500 m.

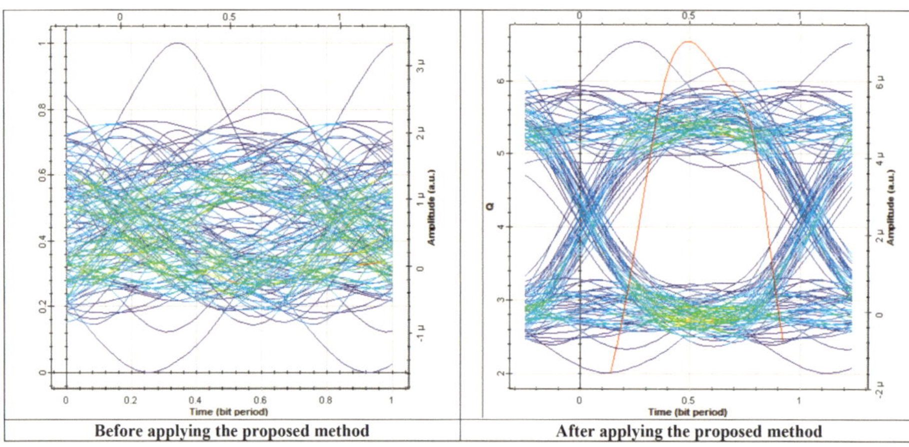

Figure 17.
Eye diagram before and after the application of the proposed method at damping 3.1 and the distance 10,000 m.

distance 10,000 m the value of the BER improved to 3.13463e-011 (compared to the value (1) before improvement) as shown in **Figure 13**.

For attenuation value 0.47 dB/km the Q factor value improved to 7.578 at the distance 6500 m (compared to the value 5.781 before improvement). At the distance 10,000 m the value of the BER improved to 6.754 (compared to the value (0) before improvement) as shown in **Figure 14**.

For attenuation value 3.1 dB / km the Q factor value improved to 19.958 at the distance 3000 m (compared to the value 6.348 before improvement). At the distance 10,000 m the value of the BER improved to 6.534 (compared to the value (0) before improvement) as shown in **Figure 15**.

The results also show an improvement in the opening of the eye after using the proposed method, for attenuation value 3.1 dB/km the eye height value improved to 4.83267e-006 at the distance 3000 m (compared to the value (0) before improvement). At the distance 10,000 m the value of the eye height improved to 2.7099e-006 (compared to the value (0) before improvement) as shown in **Figures 16** and **17**.

10. Conclusions

In this chapter, a new method was presented to improve the performance of the wireless optical link through the suggestion of adaptive power control, by modifying the power of the optical amplifier based on changing weather conditions. We obtain an acceptable BER and a good quality factor, while maintaining the minimum power required for the transmitter.The proposed system was tested on attenuation values (0.47, 3.1) dB/km and results have shown a clear improvement in the received power, bit error rate, quality factor, and eye diagram after using the proposed method.

Author details

Mazen Abdullatif
Aleppo University, Aleppo, Syria

*Address all correspondence to: mazenabdullatif87@gmail.com

References

[1] Khalighi MA, Uysal M. 2014-Survey on free space optical communication: A communication theory perspective. In: IEEE Communications Surveys & Tutorials. 4th Quart. 2014;**16**(4):2231-2258

[2] Mikolajczyk J, Bielecki Z, Bugajski M, Piotrowski J. Analysis of free space optics development. Metrology and Measurement Systems. 2017;**24**(4): 653-674

[3] Kumar A, Dhiman DK, Kumar N. Free space optical communication system under different weather conditions. IOSR Journal of Engineering (IOSRJEN). 2013;**3**(12):52-58

[4] Odeyemi KO, Owolawi PA. A mixed FSO/RF integrated satellite-high altitude platform relaying networks for multiple terrestrial users with presence of E avesdropper: A secrecy performance. Photonics. 2022;**9**:32

[5] Hamza AS, Deogun JS, Alexander DR. Wireless communication in data Centers: A survey. IEEE Communications Surveys and Tutorials. 2016;**18**:1572-1595

[6] Jurado-Navas A, Garrido-Balsells JM, Castillo-Vázquez M, García-Zambrana A, Puerta-Notario A. Converging underwater and FSO ground communication links. In: Proceedings of the 2019 Optical Fiber Communications Conference and Exhibition (OFC), San Diego, CA, USA, 3–7 March 2019. pp. 1-3. (Invited paper)

[7] Qin Y, Kishk MA, Alouini MS. Drone charging stations deployment in rural areas for better wireless coverage: Challenges and solutions. IEEE Internet of Things Magazine. 2022;**5**:148-153

[8] Kim II, McArthur B, Korevaar EJ. Comparison of laser beam propagation at 785 and 1550 nm in fog and haze for optical wireless communications. SPIE. 2001;**4214**(2):26-37

[9] Cvijetic M. Optical Transmission Systems Engineering. Boston London: Artech House, Inc.; 2004

[10] Pospiech M, Liu S. Laser Diodes an Introduction. Germany: University of Hannover; May 2004

[11] Shieh W. OFDM for Optical Communications. Elsevier; 2010

[12] Mohammed NA, El-Wakeel AS, Aly MH. Performance Evaluation of FSO Link Under NRZ-RZ Line Codes, Different Weather Conditions and Receiver Types in the Presence of Pointing Errors. The Open Electrical & Electronic Engineering Journal. 2012;**6**:28-35

[13] Desurvire E. Erbium-Doped Fiber Amplifiers – Principles and Applications. USA: John Wiley & Sons, Inc.; 1994

[14] Burgmeier J, Cords A, März R, Schäffer C, Stummer B. A black box model of EDFA's operating in WDM systems. Journal of Lightwave Technology. 1998;**16**(7):1271-1275

[15] Bastien SP, Sunak HRD, Sridhar B, Kalomiris VE. Temporal, spatial and spectral modeling of erbiumdoped fiber amplifiers. SPIE – Physic and Simulations of Optoelectronic Devices. 1992:2-11

[16] Bloom S, Korevaar E, et al. Understanding the performance of free-space optics. Journal of Optical Networking. 2003;**2**(6):178-200

[17] Chieng JLY, Hassan I. Optical wireless communication system. Journal of Engineering Science and Technology. 2016

Section 3

Free Space Optical Channel Characteristics for Future-Oriented 6G Networks

Chapter 4

Fundamental Channel Characteristic Comparison of Distinct LED Beams-Based Outdoor Free Space Optical Communications for 6G Wireless Networks

Jupeng Ding, Chih-Lin I, Jintao Wang and Jian Song

Abstract

In upcoming 6G era, free space optical wireless communications (FSOC) are viewed as promising enabling techniques for backhaul and access of next-generation wireless networks. Channel modeling and characterization is the essential composition to predict FSOC system performance and feature, especially for outdoor application. Objectively, light emitting diodes (LED) are the popular and cost-efficient optical sources for short- and middle-range outdoor FSOC links. Up to know, the concerned LED beams are assumed to follow the well-known Lambertian radiation patterns for outdoor FSOC domain, which actually ignore the diversity and effect of the non-Lambertian radiation patterns of the commercially customized LEDs. For addressing the above issue, typical LUXEON Rebel non-Lambertian optical beam is adopted to configure the distinct LED beams-based outdoor FSOC links. Numerical results illustrate that more dispersive horizontal coverage characteristic could be derived from the LUXEON Rebel non-Lambertian FSOC links.

Keywords: free-space optical communications, channel characteristics, non-Lambertian beams, outdoor scenario, 6G

1. Introduction

Thanks to the impressive advantages in ultra-high capacity, abundant optical spectrum resource, low latency, electromagnetic interference free and reliable security, free-space optical wireless communications (FSOC) are considered and investigated as one promising enabling technique for the upcoming 6G wireless networks [1–5]. Specifically, the FSOC has been consistently explored in various outdoor application scenarios including wireless backhauls, drone communications, vehicle communications, satellite communications, and emergency communications. Due to the harsh optical signal propagation condition, the actual performance of FSOC is significantly affected by the outdoor optical channel characteristics in distinct application scenarios. Up to know, most works are limited to the investigation of outdoor long range, even ultra-long range FSOC channel characterization and performance optimization, which to a large extent ignore the research and exploration of middle- or short-range FSOC, especially the promising Internet of Things (IoT) scenarios [6–10].

Unlike the popularity of laser diode (LD) in the long-range FSOC domain, the light-emitting diodes (LED) are more frequently adopted in the middle- or short-range FSOC, thanks to its low cost, safety, and ubiquity [11–16]. At the same time, it must be noted that the current LED-based FSOC works are still restricted to the Lambertian optical beam configuration, which obviously could not touch or present the link performance of FSOC using distinct non-Lambertian LED source [11–12]. For addressing this fundamental challenge, to the best of our knowledge, for the first time, the LUXEON Rebel non- Lambertian LED source beam is proposed to configure the outdoor FSOC links, as shown in **Figure 1**. Moreover, the key performance metrics are comparatively analyzed for the channel characteristics of envisioned non-Lambertian outdoor FSOC and the baseline outdoor Lambertian FSOC.

2. Channel modeling of outdoor FSO communications

In this part, the typical non-Lambertian FSOC links is explored in typical outdoor scenario, as shown in **Figure 1** for the first time.

2.1 Case 1 of benchmark Lambertian FSO communications

As the benchmark outdoor link configuration of FSOC, the Lambertian LED source is utilized at the optical transmitter. For the purpose of clarity, the respective Lambertian optical beam could be profiled by the angle-dependent radiation intensity as [1, 12]:

$$I_{\mathrm{L}}(\theta) = \frac{m_{\mathrm{L}} + 1}{2\pi} \cos^{m_{\mathrm{L}}}(\theta) \tag{1}$$

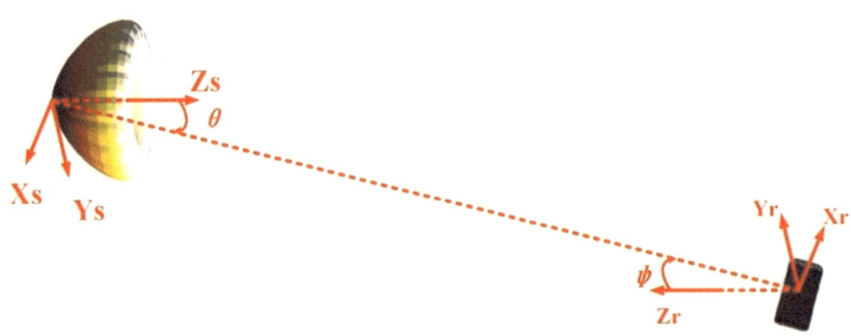

Figure 1.
Typical outdoor FSOC link horizontal configuration.

where mL is related to the order of Lambertian with half power at φ1/2, and θ is the angle of irradiance with respect to the axis normal to the transmitter surface, while the typical 3D spatial radiation pattern of Lambertian optical beam for horizontal FSOC link is shown in **Figure 2a**. Accordingly, the optical intensity channel gain $H(S^{\text{Lam}}, R)$ between the Lambertian source S^{Lam} and the optical receiver R can be expressed by [12, 14–16]:

$$H\left(S^{\text{Lam}}, R\right) = \begin{cases} \dfrac{A_R}{d_0^2} I_{\text{L}}(\theta)\cos(\psi) G_{\text{of}} G_{\text{oc}} e^{-\mu d_0}, 0 \le \psi \le FOV \\ 0, \qquad\qquad \psi > FOV \end{cases} \tag{2}$$

where AR is the physical area of the photodiode at the receiver, $d0$ is the line of sight distance from source S^{Lam} and the optical receiver R, and ψ is the angle of incidence with respect to the axis normal to the receiver surface. FOV is the field of view (FOV) of the optical receiver. Moreover, $C$of represents the optical filter gain of the receiver, Coc represents the optical concentrator gain of the receiver, and μ denotes the atmospheric attenuation coefficient, which is wavelength dependent and could be given by [14–16]:

$$\mu = \frac{3.91}{V}\left(\frac{\lambda}{550}\right)^{-q}, \tag{3}$$

where V is the atmospheric visibility, λ is the optical signal wavelength, and q is the parameter related to the visibility, expressed as [14–16]:

$$q = \begin{cases} 0.585\text{V}^{1/3}, & \text{V} < 6\text{ km} \\ 1.3, & 6\text{ km} < \text{V} < 50\text{ km} \end{cases} \tag{4}$$

In addition, replacing Eq. (1) in Eq. (2), we obtain the renewed optical intensity channel gain as:

$$H\left(S^{\text{Lam}}, R\right) = \begin{cases} \dfrac{(m_{\text{L}} + 1)A_R}{2\pi d_0^2} \cos^{m_{\text{L}}}(\theta)\cos(\psi) G_{\text{of}} G_{\text{oc}} e^{-\mu d_0}, & 0 \le \psi \le FOV \\ 0, & \psi > FOV \end{cases} \tag{5}$$

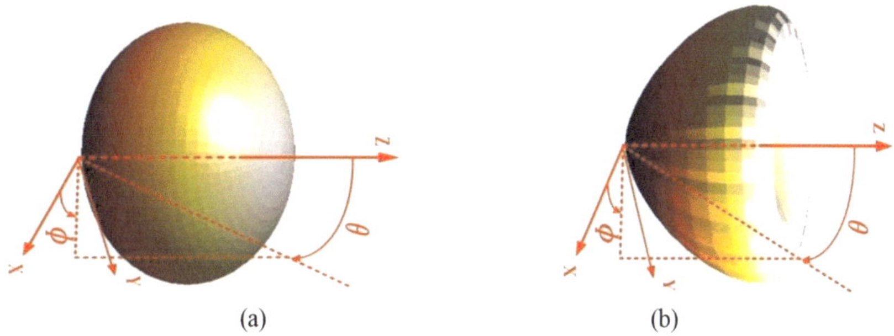

Figure 2.
3D spatial radiation patterns of (a) Lambertian optical beam and (b) typical non-Lambertian optical beam for outdoor horizontal FSOC links.

In this situation, the received optical power at the receiver could be expressed as:

$$P_r\left(S^{\mathrm{Lam}}, R\right) = \begin{cases} \dfrac{P_t(m_L+1)A_R}{2\pi d_0^2} \cos^{m_L}(\theta)\cos(\psi)G_{of}G_{oc}e^{-\mu d_0}, & 0 \le \psi \le FOV \\ 0, & \psi > FOV \end{cases} \quad (6)$$

where *Pt* is the emitted optical power at the LED transmitter and the distance between the source S^{Lam} and the optical receiver *R* can be calculated as follows:

$$d_0 = \sqrt{(X_S - X_R)^2 + (Y_S - Y_R)^2 + (Z_S - Z_R)^2} \quad (7)$$

where (*XS*, *YS*, *ZS*) denotes the coordinated position of the LED transmitter and (*XR*, *YR*, *ZR*) denotes the coordinated position of the optical receiver.

2.2 Case 2 of typical non-Lambertian FSO communications

Without loss of generality, the typical non-Lambertian radiation pattern from the commercially available LUXEON Rebel LED is adopted to configure the outdoor FSOC link for the envisioned case 2. Following the work of Refs. [11, 12], the radiant intensity of this rotational symmetric non-Lambertian type could be given as:

$$I_{NL}(\theta) = \sum_{i=1}^{2} g_{1i} \exp\left[-\ln 2\left(\frac{|\theta| - g_{2i}}{g_{3i}}\right)^2\right] \quad (8)$$

where the involved equation coefficients are set as g11 = 0.76, g21 = 0°, g31 = 29°, g12 = 1.10, g22 = 45°, g32 = 21°successively. The respective 3D spatial radiation pattern of this non-Lambertian optical beam for outdoor horizontal FSOC link is illustrated in **Figure 2b**. Apparently, like the mentioned Lambertian case, the intensity is independent of the azimuthal angle Φ which fundamentally induces the radiation pattern symmetry in the far field. Therefore, similar to the Lambertian case, the line of sight optical channel gain for this novel non-Lambertian case could be presented as:

$$H\left(S^{NL}, R\right) = \begin{cases} \dfrac{A_R}{d_0^2} I_{NL}(\theta)\cos(\psi)G_{of}G_{oc}e^{-\mu d_0}, & 0 \le \psi \le FOV \\ 0, & \psi > FOV \end{cases} \quad (9)$$

Moreover, replacing Eq. (8) in Eq. (9), the renewed optical intensity channel could be given as:

$$H\left(S^{NL}, R\right) = \begin{cases} \dfrac{\sum_{i=1}^{2} g_{1i} \exp\left[-\ln 2\left(\frac{|\theta| - g_{2i}}{g_{3i}}\right)^2\right]}{d_0^2} A_R\cos(\psi)G_{of}G_{oc}e^{-\mu d_0}, & 0 \le \psi \le FOV \\ 0, & \psi > FOV \end{cases} \quad (10)$$

Such that, for the concerned non-Lambertian configuration, the angle dependent received optical power at the receiver could be expressed as:

$$P_r(S^{NL}, R) = \begin{cases} \dfrac{P_t \sum_{i=1}^{2} g_{1i} \exp\left[-\ln 2\left(\frac{|\theta| - g_{2i}}{g_{3i}}\right)^2\right]}{d_0^2} A_R \cos(\psi) G_{of} G_{oc} e^{-\mu d_0}, & 0 \leq \psi \leq FOV \\ 0, & \psi > FOV \end{cases} \quad (11)$$

For convenience of analysis, the multipath non-line of sight components are ignored in this work, which is solid and acceptable for the most outdoor FSOC application scenarios.

3. Numerical studies

For fair comparison, the whole emitted optical power of the LUXEON Rebel LED emission patterns is normalized to 1 W. The main parameters for the following comparative simulation are included in the **Table 1**. In this Lambertian optical beam configuration case, the outdoor FSO received optical power spatial distribution experiences up to 39.88 dB intensity variation, specifically ranging from −30.40 to 9.48 dBm, as shown in **Figure 3a**.

Due to the intrinsic wider and more dispersive spatial emission characteristics of the concerned LUXEON Rebel non-Lambertian optical beam configuration, the received optical power spatial distribution ranges from −32.86 to 13.23 dBm with variation reduced to 46.09 dB. Accordingly, the horizontal coverage performance difference brought by the pattern replacement could been observed by the cumulative distribution function (CDF) in **Figure 4** as well.

For up to 80% outdoor FSO receiver positions under Lambertian optical beam configuration, the received optical power is more than −28.63 dBm, while the counterpart of the LUXEON Rebel non-Lambertian optical beam configuration is changed to −32.71 dBm, accordingly. Therefore, the discussed LUXEON Rebel non-Lambertian optical beam configuration is capable of providing more dispersive horizontal coverage at the price of 4.08 dB maximum intensity loss for 80% receiver positions. As also shown in **Figure 4**, the similar tendency and statistical characteristics difference could be identified when the emitted optical power is increased to 60 W and 90 W for the concerned Lambertian optical beam configuration and LUXEON Rebel non-Lambertian optical beam configuration.

Parameters	Value
Wavelength	570 [nm]
Visibility	20 [km]
Link length	50 [m]
Emitted optical power	30 [W]
Lambertian index	1
Detection area of receiver	0.01 m^2
Field of view	90°
Optical filter gain	1
Optical concentrator gain	1

Table 1.
Parameters for outdoor FSOC channel characteristics simulation.

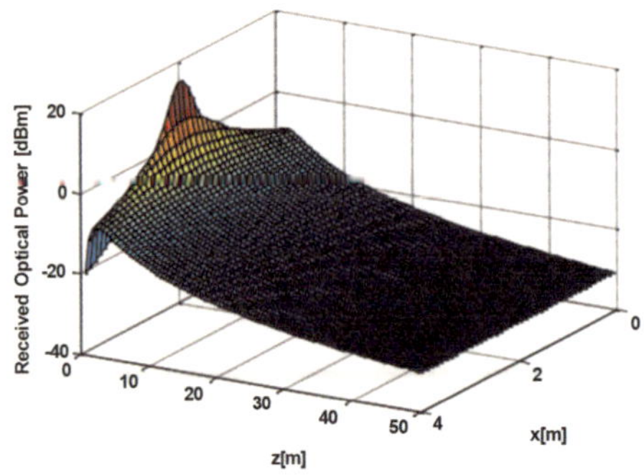

(a)

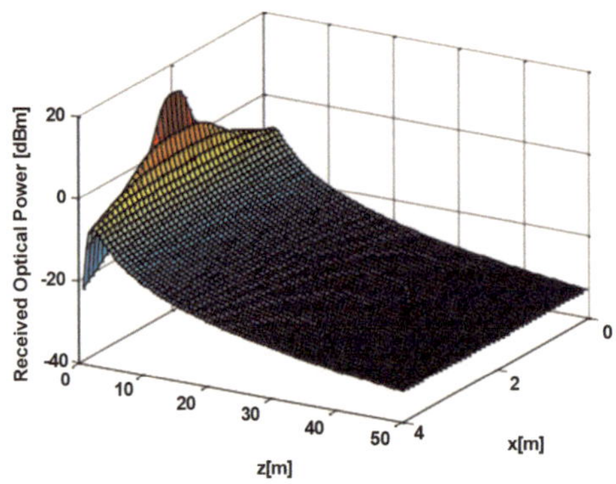

(b)

Figure 3.
Received optical power spatial distribution within horizontal plane coverage of outdoor FSOC for 6G wireless networks in the case of (a) Lambertian optical beam configuration, (b) LUXEON Rebel non-Lambertian optical beam configuration.

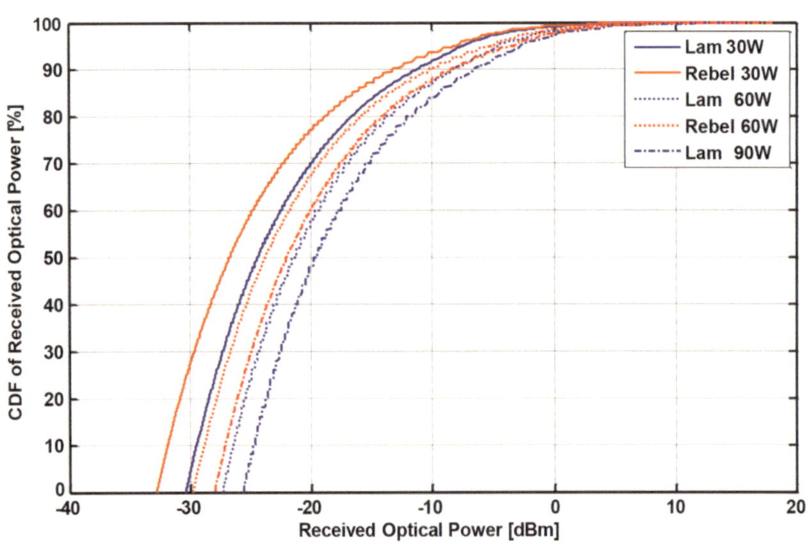

Figure 4.
CDF curve comparison of received optical power within horizontal plane coverage for two distinct optical beam configurations.

4. Conclusion

The ultra-high capacity, abundant optical spectrum resource, low latency, electromagnetic interference free, and reliable security of FSOC are quite competitive and attractive to developing 6G wireless networks. For overcoming the limitation of current LED-based outdoor FSOC technology in optical beam configuration, in this work, the relevant channel characteristics induced by the non-Lambertian LED beam are investigated in typical outdoor FSOC scenario. The respective results illustrate that more dispersive for outdoor FSOC could be introduced by the LUXEON Rebel non-Lambertian optical beam configuration, compared with the baseline Lambertian FSOC case.

Acknowledgements

This work is supported by National Natural Science Foundation of China (Grants No. 62061043), Natural Science Foundation of the Xinjiang Uygur Autonomous Region (Grants No. 2019D01C020), and Tianshan Cedar Project of Xinjiang Uygur Autonomous Region (Grants No. 202101528).

Conflict of interest

The authors declare no conflict of interest.

Author details

Jupeng Ding[1]*, Chih-Lin I[2], Jintao Wang[3] and Jian Song[3]

1 Key Laboratory of Signal Detection and Processing in Xinjiang Uygur Autonomous Region, School of Computer Science and Technology (School of Cyberspace Security), Xinjiang University, Urumqi Xinjiang, China

2 China Mobile Research Institute, Beijing, China

3 Department of Electronic Engineering, Beijing National Research Center for Information Science and Technology, Tsinghua University, Beijing, China

*Address all correspondence to: jupeng7778@163.com

References

[1] Chowdhury MZ, Hasan MK, Shahjalal M, Hossan MT, Jang YM. Optical wireless hybrid networks: Trends, opportunities, challenges, and research directions. IEEE Communications Surveys & Tutorials. 2020;**22**:930-966. DOI: 10.1109/COMST.2020.2966855

[2] Pham T, Atsushi K, Keizo I, Toshimasa U, Naokatsu. Hybrid optical wireless-mmWave: Ultra-high-speed indoor communications for beyond 5G. IEEE INFOCOM 2019 - IEEE Conference on Computer Communications Workshops (IN-FOCOM WKSHPS). Paris, France: IEEE; 2019. pp. 1003-1004. DOI: 10.1109/INFCOMW.2019.8845283

[3] Huang C, Hu S, Alexandropoulos GC, Zappone A, Yuen C, Zhang R, et al. Holographic MIMO surfaces for 6G wireless networks: Opportunities, challenges, and trends. IEEE Wireless Communications. 2020;**27**:118-125. DOI: 10.1109/MWC.001.1900534

[4] Khan M, Chakareski J. Neighbor Discovery in a Free-Space-Optical UAV Network. In: 2019 IEEE Global Communications Conference (GLOBECOM). Waikoloa, HI, USA: IEEE; 2019. pp. 1-6. DOI: 10.1109/GLOBECOM38437.2019.9014063

[5] Nakamura K, Nakagawa S, Matsubara H, Tatsui D, Seki K, Haruyama S, et al. Development of broadband telecommunications system for railways using laser technology. Electrical Engineering. 2015;**132**(5):666-674. DOI: 10.1541/ieejeiss.132.666

[6] Mori K, Terada M, Yamaguchi D, Nakamura K, Kaneko K, Teraoka F, et al. Fast handover mechanism for high data rate ground-to-train free-space optical communication transceiver for internet streaming applications. IEICE Transactions on Communications. 2016;**E99.B**:1206-1215. DOI: 10.1587/transcom.2015EBP3326

[7] Cossu G, Sturniolo A, Messa A, Scaradozzi D, Ciaramella E. Full-fledged 10Base-T ethernet underwater optical wireless communication system. IEEE Journal on Selected Areas in Communications. 2018;**36**:192-202. DOI: 10.1109/JSAC.2017.2774702

[8] Mozaffari M, Saad W, Bennis M, Debbah M. Optimal transport theory for power-efficient deployment of unmanned aerial vehicles. In: Processing of the 2016 IEEE International Conference on Communications (ICC), Kuala Lumpur, Malaysia: IEEE; 2016, pp. 1-6. DOI: 10.1109/ICC.2016.7510870

[9] Zhang Y, Yang Y, Hu B, Yu L, Hu Z. Average BER and outage probability of the ground-to-train OWC link in turbulence with rain. Optics Communications. 2017;**68**:85-90. DOI: 10.1016/j.optcom.2017.04.034

[10] Kim I, Mcarthur IB, Korevaar EJ. Comparison of laser beam propagation at 785 nm and 1550 nm in fog and haze for optical wireless communications. In: Proceedings Volume 4214, Optical Wireless Communications III; (2001). Boston, MA, United States: Event: Information Technologies; 2000. DOI: 10.1117/12.417512

[11] Moreno I, Sun CC. Modeling the radiation pattern of LEDs. Optics Express. 2008;**16**(3):1808-1819. DOI: 10.1364/OE.16.001808

[12] Ding J, Chih-Lin I, Xu Z. Indoor optical wireless channel characteristics

with distinct source radiation patterns. IEEE Photonics Journal. 2016; **8**(1):1-15. DOI: 10.1109/JPHOT. 2015.250842

[13] Nouri H, Uysal M. Adaptive MIMO FSO communication systems with spatial mode switching. Journal of Optical Communications and Networking. 2018; **10**(8):686-694. DOI: 10.1364/ JOCN.10.000686

[14] Bouanane L, Abbou MF, Abdi F, Smith K, Abid A. Power penalty analysis of IM-DD free space optical Communication Channel in the presence of turbulence. In: 2023 International Wireless Communications and Mobile Computing (IWCMC). Marrakesh, Morocco: IEEE; 2023. pp. 1424-1429. DOI: 10.1109/IWCMC58020.2023. 10182560

[15] Xin L, Xin-Yue G, Bo-Qing X. Channel modeling for outdoor visible light communication. Optical Communication Technology. 2016;**4**: 41-43. (in Chinese)

[16] Georlette V, Bette S, Brohez S, Pérez-Jiménez R, Point N, Moeyaert V. Outdoor visible light Communication Channel Modeling under smoke conditions and analogy with fog conditions. Optics. 2020;**1**(3):259-281. DOI: 10.3390/opt1030020

Chapter 5

WDM-FSO System for 5G Wireless Network Using Spatial Multiplexing in the Presence of Wireless and Optical Nonlinearities

Raman Jee and Somnath Chandra

Abstract

This chapter focuses on theoretical analysis of the transmission performance of wavelength division multiplexed (WDM) free-space optical communication (FSO) links amidst optical and wireless nonlinearities. It has been shown that, to mitigate the nonlinearities due to atmospheric turbulence and optical path, spatial multiplexing technique has been utilized, coupled with advanced modulation formats such as CS-RZ, QPSK and DSPK, and QAM. An in-depth theoretical model and simulation study have been carried out. In the presence of strong and medium turbulence, our analysis indicates that modulation formats QAM, DPSK, and QPSK perform better than normal baseband CS-RZ. The analysis further shows that using spatial multiplexing (MIMO), QAM has a 5.1 dB advantage over CS-RZ and a 1.6 dB advantage over DPSK and QPSK modulations at the FEC threshold level of 3.8×10^{-3}. The WDM-FSO system can transmit wireless signals up to a distance of 36 kilometers at the FEC threshold. Further, we have shown that a combination of the WDM-FSO system can transport wireless signals up to 36 km at the FEC threshold 3.8×10^{-3}. Specifically, it has been demonstrated that the spectrally efficient QAM modulation format achieves higher transmission efficiency at FEC error rate 3.8×10^{-3}.

Keywords: WDM (wavelength division multiplexed), FSO (free-space optical communication), MIMO (multiple input-multiple output), QAM, DPSK, QPSK and CS-RZ

1. Introduction

The swift development of 5G and 6G wireless systems, together with their impending deployment, necessitates increased system capacity, enhanced spectral efficiency, and faster data rates for both front- and back-haul networks [1–6]. The combined transfer of enormously diverse data traffic from several network planes and massively networked devices into the core network is the main difficult issue facing 5G and B5G wireless systems. This issue suggests that 5G systems must employ multiplexing, a spectrally efficient modulation format (MIMO), and techniques seamless integration with the core optical backbone transmission path. In the 5G scenario, a backhaul network needs dense cellular structures to provide data access, which can be accomplished using microwave and millimeter-wave systems [7, 8]. However, they face limitation signal due to degradation, fading, and spectrum scarcity. Uses of signal transmission through optical fiber transmission has the benefit of being transparent to wireless signals, allowing for the sub-carrier multiplexing of numerous heterogeneous wireless channels using coherent systems or optical intensity modulation (IM-DD). Consequently, a dual hybrid optical transmission combination such as optical-wireless free-space optics (FSO) for the wireless signal part, and wavelength division multiplexing (WDM) for optical path can utilize, and 5G needs its complementary strengths. High bandwidth, high capacity, and secure data transmission have recently been thoroughly examined [2, 9]. The statistical fading effect scintillation brought on by atmospheric turbulence limits FSO transmission and therefore the transmission distance. The FSO system will work in conjunction with WDM technology to get around that distance restriction. WDM technology can send combined data streams at a higher transmission rate by utilizing several optical wavelengths. Furthermore, a number of spectrally efficient modulation formats that can send greater data rates over channels with limited bandwidth can be used to increase the data rate. Modulation in formats such as QPSK, DPSK, and m-QAM provides resistance against fading and greater spectral efficiency. In the wireless path, they are usable.

Moreover, different spatial multiplexing techniques with multiple transmit and receive structures have recently been investigated experimentally and theoretically [4–6, 9, 10] in order to improve the robustness and battle the fading effect in the FSO channel. It has been demonstrated that system performance can be enhanced by using MIMO FSO in an intensity-modulated direct-detection approach. The MIMO-FSO system's BER performance as reported in [8, 11] only takes atmospheric distortions into account when analyzing FSO transmission. However, nonlinear distortions and fading effects in both wireless and optical channels must be considered when analyzing the WDM-FSO combination with MIMO spatial multiplexing.

The analysis of system performance, including bit error rate (BER), is analyzed in this chapter using a detailed analytical model that considers both atmospheric and optical nonlinearities. The robustness of these modulation formats, which include QPSK, QAM DPSK, CS-RZ, and QAM, has been evaluated.

2. Hybrid FSO-WDM system architecture

Figure 1(a) and **(b)** display the system architecture that utilizes FSO to backhaul dense cellular radio access in 5G environments along with WDM-based optical core transport. The WDM enhances the data rate to approximately 40 Gb/s by multiplexing FSO channels, and they share the same wavelength and system elements. Wireless and optical transports are utilized to their full potential in the above converged architecture. DWDM employs bidirectional data transmission and capacity multiplication, However, FSO offers greater deployment and reusability flexibility. Furthermore, the 5G scenario's radio access units (RAUs) and the FSO's wireless portion operate in separate frequency ranges and are not impacted in a comparableway by wireless routes. The deployment requirements dictate whether or not RAUs and FSO data transfer channels are compatible, and Refs. [9, 12, 13] have already demonstrated hybrid architectures for the deployment of backhaul/front-haul

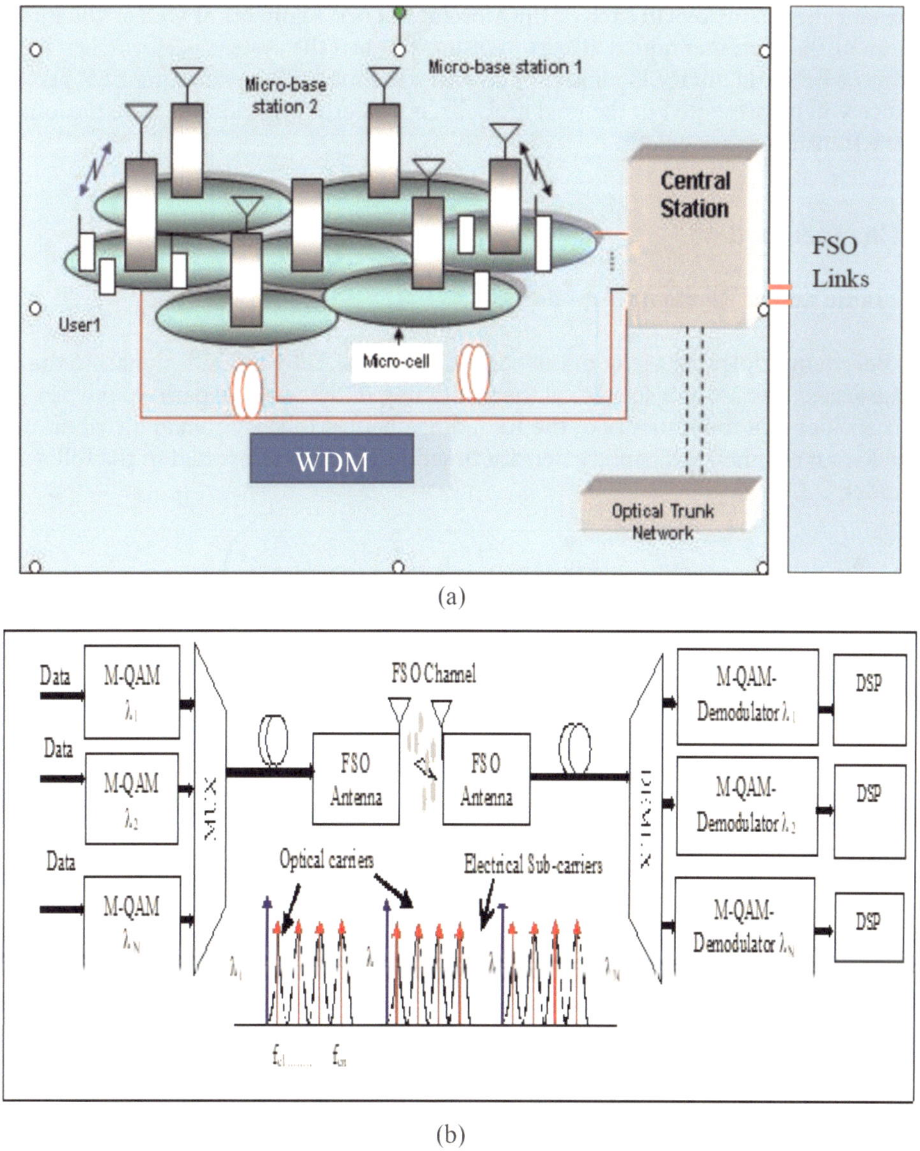

Figure 1.
(a) Schematic diagram of FSO-based access network and (b) configuration of WDM-M-QAM transmission over FSO link.

networks. In addition, optical modulation is transparent to RF signals, which allows for the use of cost-effective intensity-modulation direction detection (IM-DD) in optical transmission.

Figure 1(a) illustrates the combined FSO-WDM design for signal transmission. It is hampered by a number of nonlinear distortions in the wireless, FSO, and optical paths. The electrical signal propagating through an FSO channel experiences random intensity fluctuations, which cause scintillation. The gamma-gamma distribution can be used effectively to model medium to strong turbulence. Consequently, we must consider the distortions in each of the aforementioned segments as well as the interaction of the aforementioned effects in order to assess the system performance in terms of BER and outage likelihood. The analytical model for evaluating BER performance will be presented in the next sections, considering the nonlinear distortions in every transmission segment.

3. Channel model

3.1 Radio access RF channel model

Before multiplexing at the micro base stations, the individual RF signals in the radio access link are subjected to statistical fading in the wireless path. In our study, we consider that the statistics of the RF fading channel follow a Rician distribution. The Rician channel's probability density function (PDF) is expressed in the following manner:

$$F_{RF}(\gamma) = (1+k)\exp\left(-k-\left(\frac{\gamma}{SNR_{RF}}(1+k)\right)\right). \times I_0\left(2\sqrt{\left(\left(\frac{\gamma}{SNR_{RF}}\right)(K+K^2)\right)}\right)d\gamma \quad (1)$$

I would like to know where I_0 the zeroth order Bessel function of the first kind is located. A Rayleigh fading environment is created when $K = 0$ is used for strong fading.

3.2 Model for FSO channel

In LOS mode, the FSO link works with the pointing error being the total distance between the beam weight center and the receiver aperture center. To obtain PDF of the pointing error *hp*, the model described in [14, 15] can be used, as per the model in [8, 12, 16]. According to this model, attenuation resulting from geometric spread with radial displacement r from the detector origin can be roughly represented as follows on the receiver plane at distance L from the transmitter if we consider a circular detection aperture with radius a and a Gaussian spatial intensity profile of beam waist:

$$h_p(r,L) = A_0 \exp\left[-\frac{2r^2}{w_{l_{eq}}^2}\right] \quad (2)$$

The fraction of power collected by the detector is called hp(.), $v = \sqrt{\pi r}/\sqrt{2w_l}$, $w_{l_{eq}}^2 = w_l^2\sqrt{rerf(v)}/2v\exp(-v^2)$ is the equivalent beam width, $A = [erf(v)]^2$, erf(.) is the error function, and is the percentage of the gathered power at r = 0. In contrast, the radial displacement r has a Rayleigh distribution. After that, the *hp* PDF is provided by:

$$f_{h_p}(h_p) = \frac{\gamma^2}{A_0^{\gamma^2}} h_p^{\gamma^2-1} 0 \le h_p \le A_0 \quad (3)$$

Where $\gamma = w_{zeq}/2\sigma_s$ is the ratio between the equivalent beam radius and the standard deviation of the pointing error displacement at the receiver. The nth moment of ha is defined as $E\left[h_p^n\right] = \int_0^{A_0} h_p^n f_{h_p}(h_p)dh_p$ which can be obtained as

$$E\left[h_p^n\right] = \frac{A_0^n \gamma^2}{n+\gamma^2} \tag{4}$$

3.3 Model for atmospheric turbulence

The signal's degradation is caused by the statistical random fluctuations caused by strong and weak turbulence in the atmosphere.

The gamma-gamma model is a model that is commonly employed and can handle both weak and strong turbulence regimes. The PDF that has been provided contains the received irradiance;

$$p(I) = \frac{2(\mathrm{ab})^{\frac{a+b}{2}}}{\Gamma(\mathrm{a})\Gamma(\mathrm{b})} I^{\frac{a+b-2}{2}} K_{a-b}\left(2\sqrt{abI}\right), \mathrm{I} > 0 \tag{5}$$

where x is the normalized signal intensity scintillation, $\Gamma(.)$ is gamma function, and $Kn(.)$ is the modified Bessel function.

$$\begin{aligned} a &= \left[\exp\left\{\frac{0.49\sigma^2}{\left(1+0.18\mathrm{d}^2+0.56\sigma^{12/5}\right)^{7/6}}\right\} - 1\right]^{-1} \\ b &= \left[\exp\left\{\frac{0.51\sigma^2\left(1+0.69\sigma^{12/5}\right)^{-5/6}}{1+0.9d^2+0.62d^2\sigma^{12/5}}\right\} - 1\right]^{-1} \end{aligned} \tag{6}$$

$\sigma^2 = 0.5C_n^2 k^{7/6} L^{11/6}$ and $d = \sqrt{\left(\pi D^2/2\lambda L\right)}$, $D(m)$ is the diameter of the receiver collecting lens aperture, $\lambda(m)$ is the wavelength, and $L(m)$ is the link distance, and the parameter responsible for the refractive index structure is $C_n^2\left(m^{-2/3}\right)$. The scintillation index S.I can be expressed in terms of a and b as follows.

$$S.I = \frac{1}{a} + \frac{1}{b} + \frac{1}{ab} \tag{7}$$

The PDF can be used to express the combined effects of atmospheric turbulent and the pointer error.

$$f_h(h) = \int f_{h_p}(h_p).p(I)dI \tag{8}$$

Following the approach in [14, 17, 18], the combined conditional PDF is represented as;

$$f_h(h) = \frac{ab\phi}{h_{l2}A_0\Gamma(a)\Gamma(b)} G_{1.3}^{3.0}\left[\frac{ab}{A_0 h_{l2}} h \middle| \begin{matrix} \phi^2 \\ \phi^2-1, a-1, b-1 \end{matrix}\right] \tag{9}$$

where G (.) can I find the Meijer G function.

3.4 Combined model of radio access RF and FSO

Our belief is that there is no statistical correlation between the fading in the radio access path and the atmospheric turbulence in the FSO link. Conditional PDFs can be used to represent the channel state to assess system performance,

$$P(\psi) = F_{RF}(\gamma) \cdot f_h(h) \tag{10}$$

3.5 Distortion due to optical nonlinearities

The signal that has faded due to both RF fading and FSO atmospheric turbulence is sent through optical fiber. Different optical carriers modulate the intensity of the faded signals, which are then multiplexed to form a WDM signal. The SSMF is used to transmit each WDM signal through intensity-modulated direct detection (IM-DD). Evaluation of the system performance of the WDM-FSO involves evaluating the effects of different optical nonlinearities encountered by the signal through the SMF. After being converted to the optical carrier frequency (FC), the WDM signal for N channels can be written as:

$$S(t) = \sum_{i=0}^{N-1} s_i(t) = \sum_{i=0}^{N-1} u_i(t) \cdot \exp[j\omega_i(t) + \phi_i(t)] \tag{11}$$

The optical power output from the laser diode is proportional to the modulating signals as

$$P(t) = P_t\left(1 + \sum_{i=0}^{N-1} m_i s(t) + a_3\left[\sum_{i=0}^{N-1} m_i s(t)\right]^3\right) \tag{12}$$

The average transmitted optical power, a3, and mi are the average OMI per channel, which is Pt. The equations govern the small signal intensity modulation (IM) and phase modulation (PM) as they propagate through the fiber, taking into account both chromatic dispersion and SPM effect.

$$\frac{\partial \tilde{p}_N}{\partial z} = \beta_2 \omega^2 P_0 \tilde{\phi} \tag{13}$$

$$\frac{\partial \tilde{\phi}}{\partial z} = -\left[\frac{\beta_2 \omega^2}{4P_0} + \gamma \exp(-\alpha z)\right] \tilde{p}_N \tag{14}$$

where $\tilde{p}_N = \tilde{p}_{N(z,\omega)}$, $\tilde{\phi} = \tilde{\phi}(z,\omega)$ is the frequency of angular modulation, t is the time in a frame moving at the fiber input, and z is the distance. α is the coefficient of power attenuation, γ is the nonlinearity coefficient owing to, and β_2 is the dispersion coefficient to nonlinear Kerr effect. where $\tilde{p}_N = \tilde{p}_{N(z,\,\omega)}$ and $\tilde{\phi} = \tilde{\phi}(z,\omega)$ are the Fourier transforms of the normalized IM and PM terms $p_N(z,t)$ and $\phi(z,t)$, respectively $\omega = 2\pi f$. The transmission of signals may experience IM-IM and PM-IM conversions due to small variations in the modulated optical signal caused by fiber dispersion and non-linear effects.

Fiber transfer function that incorporates PM-IM and IM-IM effects can be characterized by [17, 19, 20],

$$H_F(z,\omega) = C_{IM-IM}(z,\omega) + \frac{H_{PM}(\omega)}{2} C_{PM-IM}(z,\omega) \tag{15}$$

where

$$C_{PM-IM} = -\frac{2\alpha\beta}{\nu\gamma P_0} \sin\left(\frac{\alpha\nu z}{2}\right) = 2\sin\left(\frac{\beta_2\omega^2 z}{2}\right) \tag{16}$$

and

$$C_{IM-IM} = \cos\left(\frac{\alpha\nu z}{2}\right) = \cos\left(\frac{\beta_2\omega^2 z}{2}\right) \tag{17}$$

$$H_{PM}(\omega) = 2P_0\tilde{\phi}_{in}(\omega)/\tilde{p}_{in}(\omega) \tag{18}$$

For directly modulated laser,

$$H_{PM}(\omega) = \alpha_c\left(1 - jf_c/f\right) \tag{19}$$

4. Derivation of BER expression

In this section, we present analytical BER expressions that take into account both optical nonlinearities and the effects of RF fading and atmospheric turbulence on the FSO path. The BER will be presented without any diversity first. The analytical expressions for BER with spatial multiplexing will be introduced in the future.

4.1 BER without spatial multiplexing in the single input-single output (SISO) scenario

The signal to noise ratio (SNR) and distortion ratio (Distortion) ratio in the optical channel should be evaluated before evaluating the BER. After transmission through the optical path, the faded signal from the RF-FSO path becomes worse because of optical fiber dispersion and the nonlinear Kerr effect. Clipping noise is occurring in the signal due to the rapid amplitude fluctuations of the multiplexed RF signal that exceed the laser liner threshold. The equation above shows that the required signal experiences multiple distortions caused by fiber dispersion and phase modulation, such as SPM. The signal that was received after the fiber transmission is a description.

$$y(t) = h_{fiber}(t) \otimes s(t) + n_d(t) + n_{AWGN}(t) + n_I(t) \tag{20}$$

On each side, apply a Fourier transform to

$$Y(\omega) = H_{fiber}(\omega) \cdot S(\omega) + N_d(\omega) + N_G(\omega) + N_I(\omega) \tag{21}$$

After a few mathematical steps, using Eqs. (14) and (15), the received signal-to-noise ratio can be evaluated as

$$SNDR = \overline{\gamma} = \frac{\alpha P_{RX}}{\sigma_d^2 + \sigma_{NLI}^2 + \sigma_G^2 + \sigma_I^2} \tag{22}$$

$$\text{where } \sigma_d^2 = m_i^2 \left[f_m\left(D(\omega)\eta\lambda^2/c\right)\right]^2 \tag{23}$$

$$\text{and } \sigma_{NLI}^2 = \left[\frac{1}{2} m \cdot \frac{\lambda^2}{2\pi c} \cdot D \cdot k \cdot n_2 \cdot \frac{P_0}{A_{eff}} \cdot (\omega)^2 \cdot z^2(L)\right]^2 \cdot N_{CSO} \tag{24}$$

N_{CSO} is the number of second-order signal-interference beating terms. The clipping noise can model as Poisson process and can be evaluated using [16]

$$\sigma_{clip}^2 = \sigma_I^2 = \frac{4\overline{\tau}}{3} \pi^{-3/2} m_{total}^5 \exp\left(\; 1/\mathrm{m}_{total}^2\right)\left(\rho \mathrm{P_r}\right)^2 \tag{25}$$

Gaussian noise arises from the ASE noise generated by the optical amplifier in the transmission path. By using, it is possible to evaluate the BER of an optical signal transmitted in the presence of Gaussian noise and optical nonlinear distortion for different modulations.

$$p_{CS-RZ} = Q\left(\sqrt{I^2\overline{\gamma}/2}\right); \text{CS-RZ case} \tag{26}$$

$$p_{DPSK} = \frac{1}{2} \exp\left(-I^2\overline{\gamma}\right); \text{DPSK case} \tag{27}$$

$$P_{s,16-QAM,Av} = \frac{\left(1 - M^{-1/2}\right)}{N \log_2(M)} \times \sum_{i=0}^{N-1} \int_0^{\infty} \left[2erfc\left(\sqrt{\frac{\overline{\gamma}(I)}{2(M-1)}}\right) - \left(1 - \sqrt{M^{-1}}\right) erfc^2\left(\sqrt{\frac{\overline{\gamma}(I)}{2(M-1)}}\right)\right] f_1(I) dI; \text{QAM-case} \tag{28}$$

To evaluate the bit error rate (BER) in the free-space optical communication (FSO), the BERs mentioned in (26)–(28) must now be averaged across the can be PDF of the gamma-gamma log-normal channel and expressed as

$$p_e = \int_0^{\infty} p\left(I^2\overline{\gamma}\right) f(I) dI \tag{29}$$

where $p\left(I^2\overline{\gamma}\right)$ is BER of the optical signal transmitted in the additive white Gaussian noise (AWGN) channel and $\overline{\gamma}$ the average ratio of electrical signal-to-noise are represented respectively.

4.2 BER with spatial multiplexing (multi input-multiple output-MIMO case)

We are presently concentrating on FSO links that have a variety of spatial features, such as M transmit and N receive apertures. The M N routes are where the laser output signals are transmitted. The signal received by the nth receiver is given by.

$$r_n = \eta u_n \sum_{m=1}^{M} I_{mn} + g_n; n = 1, 2, \dots \dots \dots \dots N; u_n \in 0, 1 \quad (30)$$

Let us assume that the EGC technique is used to combine optical signals with equal gain factors on each branch when receiving signals are combined. The receiver's output can be described as

$$r = \sum_{n=1}^{N} r_n = \eta u_n \sum_{n=1}^{N} \sum_{m=1}^{M} I_{mn} + g_n \quad (31)$$

The instantaneous electrical signal-to-noise ratio (SNR) is a term that can be defined for an EGC receiver as follows:

$$r_{EGC} = \left(\sum_{m=1}^{M} \sum_{n=1}^{N} \sqrt{\gamma_{mn}} \right)^2 \quad (32)$$

In the event that both transmit and receive diversity are applied, the most effective decision metric for on-off keying is given by:

$$p(r|1, I_{mn}) = \frac{\exp\left(-\frac{1}{2\sigma_g^2} \sum_{n=1}^{N} r_n^2\right)}{\left(2\pi\sigma_g^2\right)^{\frac{N}{2}}} \quad (33)$$

$$p(r|0, I_{mn}) = \frac{\exp\left(-\frac{1}{2\sigma_g^2} \sum_{n=1}^{N} \left(r_n - \eta \sum_{m=1}^{M} I_{mn}\right)^2\right)}{\left(2\pi\sigma_g^2\right)^{\frac{N}{2}}} \quad (34)$$

Assuming 0, 1 equiprobable, $p_e(o|I_{mn}) = p_e(1|I_{mn})$ and averaging over the fading coefficients, we obtain p_e as,

$$[P_e]_{FSO-Optical} = \int_I f_I(I) Q\left(\frac{\eta I_0}{2MN\sigma_g} \sqrt{\sum_{n=1}^{N} \left(\sum_{m=1}^{M} e^{2x_{mn}} \right)^2} \right) dI \quad (35)$$

Eqs. (26)–(28) are now available for numerical evaluation of the BER performance of different modulation formats with SDRT in the gamma-gamma distributed turbulent channel.

4.3 Spatial multiplexing (multiple input-multiple output-MIMO case) and RF fading can be combined to form BER

The signals on the RF access path experience fading, which we have already mentioned and modeled using Rician fading as described in Eq. (10), and the overall

BER can be evaluated by averaging *[Pe]FSO-Optical* over Rician fading and expressed as,

$$\bar{P}_e = \int_0^\infty [P_e]_{FSO-Optical} P_\psi d\psi \tag{36}$$

5. Results and discussions

BER expressions, described in Section 4, were used to perform numerical analysis and simulation of the Hybrid WDM-FSO system with and without spatial diversity. The OPTIWAVE platform and Monte-Carlo method were utilized to carry out the system simulation, with the Monte-Carlo method being used. The system parameters used for simulations are as given in **Table 1**. The received spectrum of a four-channel WDM optical signal paired with a faded radio frequency signal in the radio access and the FSO path over a transmission distance of L = 36 km is shown in **Figure 2(a)** and **(b)**. In **Figure 2(c)**, **(d)**, **(e)**, and **(f)**, the eye diagrams for 16-QAM, DPSK, CS-RZ, and QPSK are displayed in their respective eye diagrams. The wireless signal can be transported by the combination of WDM and FSO, as demonstrated.

Figure 3 displays the BER performance of three modulation formats: (i) QAM with EGC receiver diversity, (ii) QPSK, and (iii) CS-RZ. The performance of the above modulation formats can be improved by the use of diversity technique, as shown in **Figure 3**. Using diversity at the FEC threshold of 3.8×10^{-3}, CS-RZ achieves a

Parameters	Description	Refs.	Proposed work
(Rb)	Bit rate	[15, 16, 18, 21]	1.25-10 Gb/s
(λ)	Optical wavelength	[12, 16]	1550 nm
α	Fiber attenuation	[15, 18, 21]	0.2 dB per km
	Length of FSO link	[15, 18, 21, 22]	0.6 km
	Atmospheric attenuation	[12, 16]	0.2 dB per km
L	Distance of fiber path	[12, 16]	20–60 km
Sth	Density of thermal noise	[15, 18, 21]	16 pA/sqrt(Hz)
ρ	Photodiode responsivity,	[15, 18, 21]	0.8 A/W
D	SSMF parameters	[15, 18, 21, 23, 24]	17 ps/(km nm)
S	SSMF parameters	[15, 18, 21]	0.0578 ps/nm^2 km
α	SSMF parameters	[15, 18, 21]	0.2 (dB/km)
γ	Nonlinear coefficient	[15, 16, 18, 21]	1.31 (1/(W km),
(OMI)	Optical modulation index	[15, 16, 18, 21, 24]	0.7
K	Rician factor for wireless channel	[15, 16, 18, 21, 23, 25]	4

Table 1.
System parameters used for the proposed work.

performance gain of 2.1 dB, as shown in **Figure 3(a)**. For (i) QAM and (ii) QPSK, the diversity gain is 5.1 dB and 2.3 dB, respectively, at the same BER level. This is because QAM functions better under medium-turbulence atmospheric scintillation

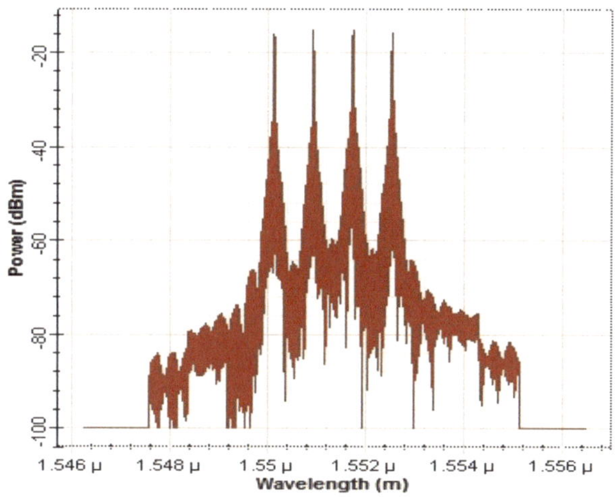

(a)

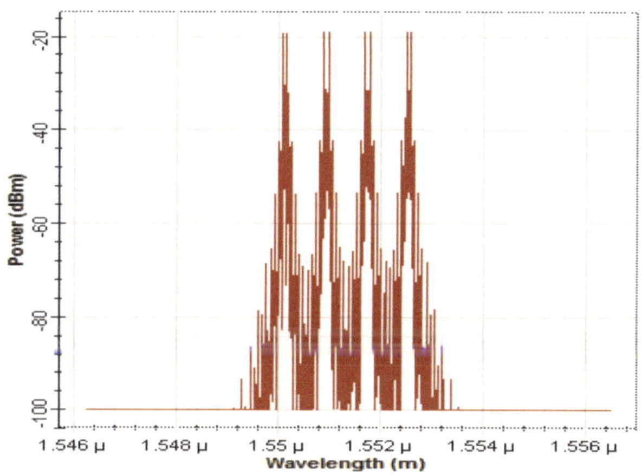

(b)

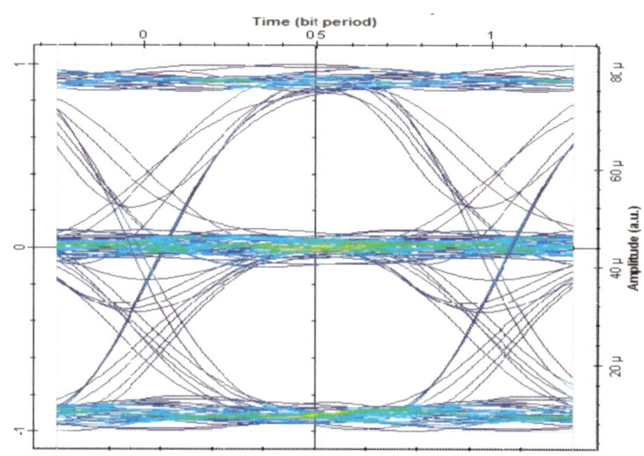

(c)

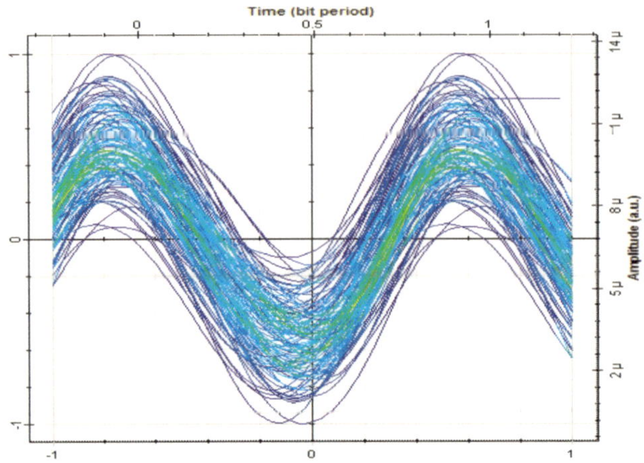

(d)

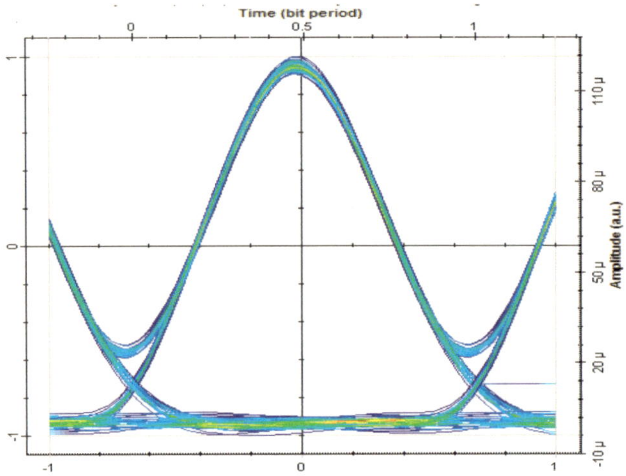

(e)

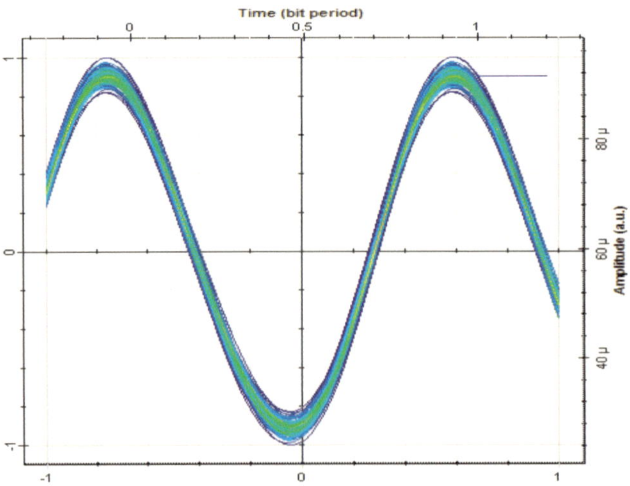

(f)

Figure 2.
(a) 16-QAM Rx optical spectrum, (b) DPSK Rx optical spectrum, (c) 16-QAM eye diagram, (d) DPSK eye diagram, (e) CS-RZ eye diagram, and (f) QPSK eye diagram.

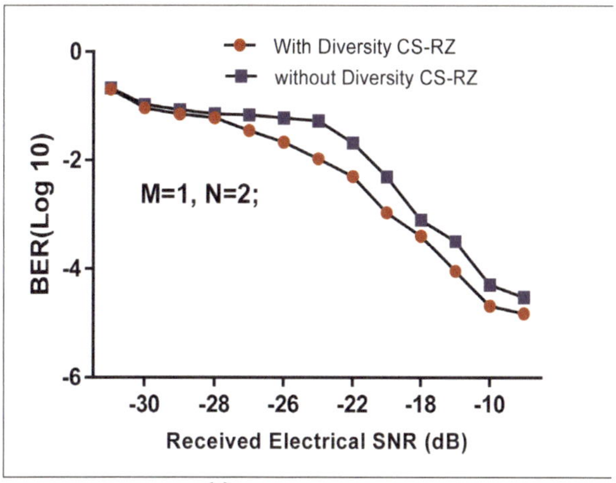

(a)

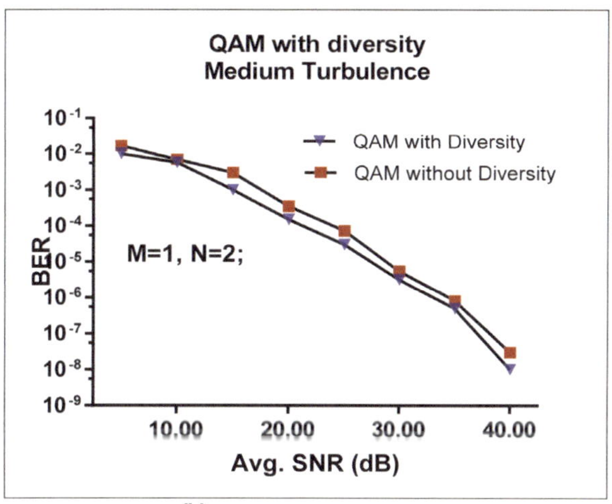

(b)

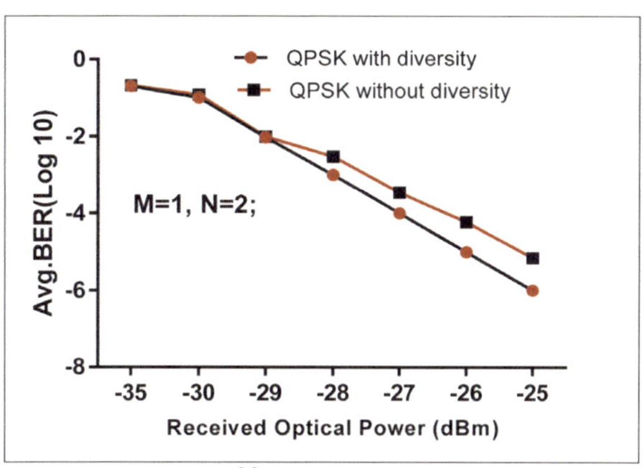

(c)

Figure 3.
SNR vs. BER for mild turbulence with receive diversity for (a) CS-RZ, (b) 16-QAM and (c) QPSK.

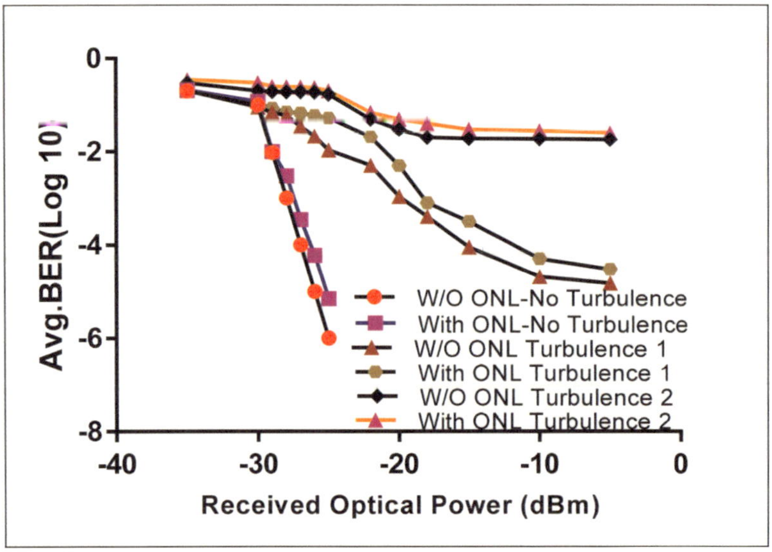

Figure 4.
QAM performance under different turbulence condition.

and has a higher spectral efficiency than the traditional CS-RZ-based QQK modulation. **Figure 4** shows the performance of QAM signals in the presence of optical nonlinearities under different turbulence conditions. It demonstrates that BER of 10^{-2} may be reached even in very turbulent environments, guaranteeing little QOS. However, the system's transmission performance is greatly impacted by the effect of optical nonlinearities. **Figure 5** shows the combined impacts of air turbulence, optical nonlinearities, and RF fading. **Figure 5(a)** and **(c)** show what has been noticed at due to the combined effects of optical nonlinearities, RF, and atmospheric fading, the QAM and QPSK signals degrade by 1.7 and 1.9 dB, respectively, to an FEC threshold of 3.8×10^{-3}. This demonstrates that the QAM signal is more resilient to the combined effects of fading and optical nonlinearities. **Figure 6** shows the overall system performance of the WDM-FSO system coupled with RF access. In this case, a four-channel WDM system has been studied. **Figure 6** shows the BER performance for the poorest channels, channels 1 and 4, in comparison to OSNR. The accompanying image illustrates how the employment of spatial diversity strategy in conjunction with WDM can achieve BER 3.8×10^{-3} at a transmission distance of 36 km.

The performance evaluation of various modulation schemes including the line-coded CS-RZ system in terms of system penalty is summarized in **Tables 2** and **3**. It has been observed that, the line-coded CS-RZ scheme has suffered a system penalty of 4 dB in the presence of turbulence of the FSO link at FEC threshold at 1.0×10^{-4}. The QPSK system has suffered less system penalty of 0.5 dB due to robustness of modulation in the presence of link turbulence. However, it has been observed that 16-QAM has suffered slightly higher system penalty of 1.2 dB. This comes at the cost of higher spectral efficiency and higher symbol rate. The DPSK has been observed to have system penalty of 2 dB. The four-channel WDM-FSO as proposed in our study suffers a system penalty of 0.4 dB as demonstrated in **Figure 6**.

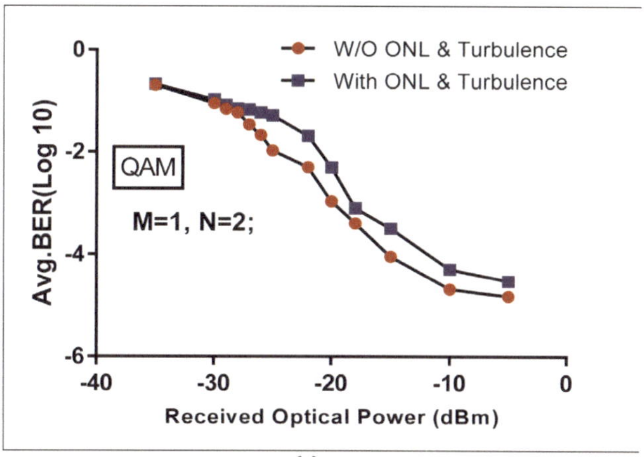

(a)

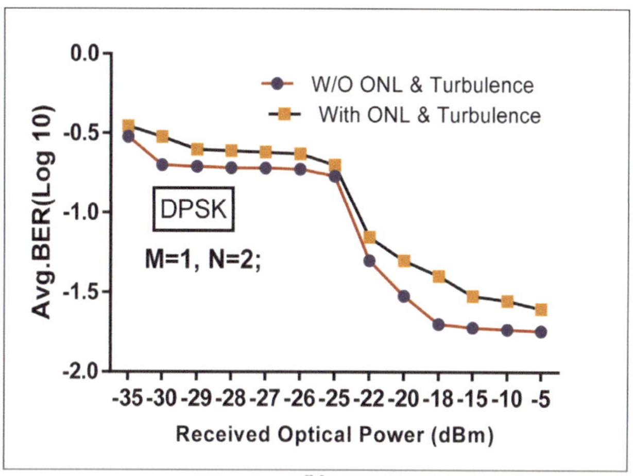

(b)

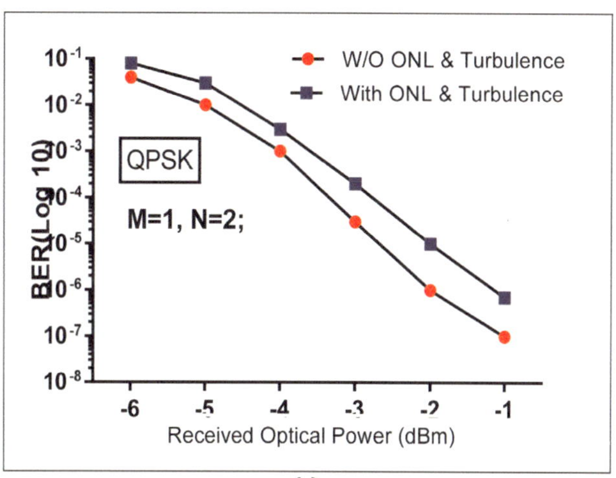

(c)

Figure 5.
(a) 16-QAM received optical power vs. BER under ONL and turbulence conditions DPSK (b) and QPSK (c).

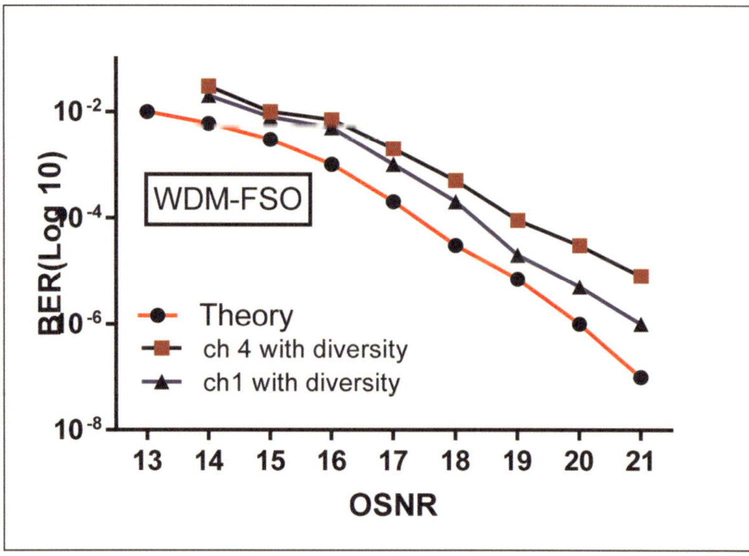

Figure 6.
OSNR versus BER can be represented by different WDM channels through WDM-FSO.

	1	2	3	4
Technique used	m-QAM	QPSK	CE-RZ	DPSK
Data rate	10 Gb/s	10 Gb/s	10 Gb/s	10 Gb/s
Net capacity	40 Gb/s	20 Gb/s	10 Gb/s	20 Gb/s
Max. FSO link range reported	600 m	800 m	1000 m	500 m
Efficiency of Quantum	0.8	0.8	0.8	0.8
Channels no.	4	4	1	4
Target BER	10^{-4}	10^{-4}	10^{-4}	10^{-4}

Table 2.
Performance comparison of hybrid WDM-FSO optical fiber link parameters.

S. no	Modulation schemes	System penalty SNR (dB)
1	CS-RZ	4.1
2	DPSK	3.0
3	QPSK	0.5
4	16-QAM	1.2

Table 3.
Performance comparison of hybrid WDM-FSO optical fiber link in terms of system with turbulence.

6. Conclusions

In this study, we have investigated and analyzed the performance of a hybrid WDM-FSO system through detailed theoretical modeling and simulation, considering the interplaying effects of RF fading, atmospheric turbulence, and optical nonlinearities. Despite the possibility of optical nonlinearities causing additional degradation effects, the optimal combination of EGC spatial multiplexing and WDM can lead to desirable system performance at FEC BER 3.8×10^{-3}. The system penalties for OSNR degradation in various line coding and modulation formats have been investigated. We found that 16 QAM offers higher spectral efficiency and bit rate, but marginal degradation under turbulent FSO conditions. This enables the WDM-FSO combination to transport wireless signals up to 36 km. To obtain higher bit rate wireless access, the analytical model mentioned above can be further extended with various other diversity techniques and DWDM.

Acknowledgements

The authors wish to thank the Ministry of Electronics and Information Technology (MeitY), Government of India, New Delhi for providing infrastructure support for conducting the above study.

Author details

Raman Jee* and Somnath Chandra
Ministry of Electronics and Information Technology, Government of India, New Delhi, India

*Address all correspondence to: ramanj35@gmail.com

References

[1] Pham AT et al. Hybrid Free-Space Optical/Millimeter-Wave Architecture for 5G Cellular Backhaul Networks. (Invited Paper)

[2] Aladeloba AO et al. WDM FSO network with turbulence-accentuated interchannel crosstalk. Journal of Optical Communications and Networking. 2013; **5**(6):641-651

[3] Bekkali A et al. Transmission analysis of OFDM-based wireless services over turbulent radio-on-FSO links modeled by gamma–gamma distribution. IEEE Photonics Journal. 2010;**2**(3):510-520

[4] Qun Shi "Performance limits on M-QAM transmission in hybrid multi-channel AM/QAM fiber optic systems" IEEE Photonics Technology Letters,1993,5(12) pp. 1452-1455.

[5] Ciaramella E et al. 1.28 Terabit/s (32x40Gbit/s) WDM transmission over adounle-pass free space optical link. OSA/OFC/NFOEC 2009

[6] Trinh PV et al. Optical amplify-and-forward multihop WDM/FSO for all-optical access networks. In: 2014 9th International Symposium on Communication Systems, Networks & Digital Signal Processing (CSNDSP). Manchester, UK: IEEE; pp 1106-1111. DOI: 10.1109/CSNDSP.2014.6923995

[7] Trung HD et al. Performance analysis of MIMO/FSO systems using SC-QAM signaling over atmospheric turbulence channels. IEICE Transactions. 2014; **E97-A**(1)

[8] Vu BT, Thang TC, et al. BER analysis of MIMO/FSO systems using rectangular QAM over gamma-gamma channels. ICUFN. Da Nang, Vietnam: IEEE; 2013. pp. 306-309. DOI: 10.1109/ICUFN.2013. 6614831

[9] Tsai W-S et al. A 50-m/320-Gb/s DWDM FSO communication with a focal scheme. IEEE Photonics Journal. 2016;**8**(3)

[10] Chen WH, Way WI. Multi-Channel single sideband SCM/DWDM transmission system. Journal of Lightwave Technology. 2004;**22**(7): 1679-1687

[11] Mohammad Navidpour S et al. BER performance of free-space optical transmission with spatial diversity. IEEE Transactions on Wireless Communications. 2007;**6**(8)

[12] Wang Z et al. Performance comparison of different modulation formats over free-space optical (FSO) turbulence links with space diversity reception technique. IEEE Photonics Journal. 2009;**1**(6)

[13] Liang P et al. Cost -efficient hybrid WDM-MDM-Ro-FSO system for broadband Services in Hospitals. Frontiers in Physics. 2021;**9**:1-7

[14] Touati A et al. On the effects of combined atmospheric fading and misalignment on the hybrid FSO/RF transmission. Journal of Optical Communications and Networking. 2016;**8**(10)

[15] Zhang J et al. Fiber-wireless integrated mobile backhaul network based on a hybrid millimetre-wave and free-space-optics architecture with an adaptive diversity combining technique. Optics Letters. 2016;**41**(9):1909-1912

[16] Jee R, Chandra S. Performance analysis of WDM-free-space optical transmission system with M-QAM Modulation under atmospheric and optical nonlinearities. In: 2015

International Conference on Advances in Computing, Communications and Informatics (ICACCI). Bhubaneswar, India: IEEE; 2015. pp. 41-44. DOI: 10.1109/ICMOCE.2015.7489686

[17] Ramos F et al. Frequency transfer function of dispersive and nonlinear single-mode optical fibers in microwave optical systems. IEEE Photonics Technology Letters. 2000;**12**(5):549-551

[18] Hai-Han L et al. Bi-directional fiber-FSO-5G MMW/5G new radio sub-THz convergence. Journal of Lightwave Technology. 2021;**39**(22):7179-7190

[19] Zhang R et al. An ultra-reliable MMW/FSO A-RoF system based on coordinated mapping and combining technique for 5G and beyond mobile fronthaul. Journal of Lightwave Technology. 2018;**36**(20):4952-4959

[20] Huang L et al. Unified performance analysis of hybrid FSO/RF system with diversity combining. Journal of Lightwave Technology. 2020;**38**(24): 6788-6800

[21] Shi Y et al. Enriching capacity and transmission of hybrid WDM-FSO link for 5G mobility. Photonics. 2023;**10**(121): 1-12

[22] Tirmizi SBR et al. Hybrid satellite-terrestrial networks toward 6G: Key technologies and open issues. Sensors. 2022;**22**:8544

[23] Li C-Y et al. A flexible bidirectional fiber-FSO-5G wireless convergent system. Journal of Lightwave Technology;**2021, 39**(5):1296-1305

[24] Bekkali A et al. Free-space optical communication systems for B5G/6G networks. In: 26th Optoelectronics and Communications Conference (OECC). OSA Technical Digest (Optica Publishing Group, 2021) paper W1A.1. 2021. DOI: 10.1364/OECC.2021.W1A.1

[25] AlGhadhban A. F4Tele: FSO for Data Center Network Management and Packet Telemetry. arXiv:2006.07419v1 [cs.NI]. 2020

Section 4

Antenna Systems for Future-Oriented 6G Networks

Chapter 6

Advancements in Antenna Systems for B5G and 6G Applications

Seyed Ramin Emadian

Abstract

This chapter explores the fundamental concepts of antenna design critical for beyond 5G (B5G) and anticipated 6G technologies. It delves into various antenna types, including microstrip antennas, metamaterial-based designs, reconfigurable antennas, phased array antenna, and lens antennas, highlighting their role in achieving the ultra-high data rates, ultra-low latency, and massive device connectivity envisioned for B5G and 6G networks. The discussion covers key aspects such as beamforming, beam steering, and pattern reconfigurability of antenna arrays, along with interference mitigation strategies. By understanding these diverse antenna systems, researchers and engineers can contribute to shaping the future of wireless communication in the B5G and 6G era.

Keywords: microstrip antennas, B5G and 6G era, phased array antenna, reconfigurable characteristics, millimeter-wave technology, massive MIMO systems

1. Introduction

In the ever-evolving landscape of wireless communications, the journey from 1G to 4G has revolutionized how we connect, communicate, and navigate the digital realm. Now, as we stand on the precipice of the Beyond 5G (B5G) era and eagerly anticipate the advent of 6G technologies, the demand for groundbreaking antenna systems takes center stage. This chapter embarks on a comprehensive exploration of the intricate world of antenna design, where innovation converges with necessity to unlock the full potential of these next-generation networks.

Wireless communication has become an integral part of our daily lives, shaping how we access information, interact with our surroundings, and engage with a multitude of devices. The progression from 1G's voice-centric systems to the data-centric networks of 4G has been marked by unprecedented speed and connectivity [1]. However, as we gaze into the future, the limitations of existing technologies become apparent. Enter B5G and 6G, promising not just incremental improvements but a seismic shift in the capabilities of wireless communication. **Figure 1** shows the evolutionary growth of wireless technology in terms of data transmission capacity and applications.

At the heart of this transformative evolution lies the antenna, a technological linchpin that translates digital signals into the physical world and vice versa. Antennas are not mere conduits; they are the gatekeepers enabling the seamless flow of information

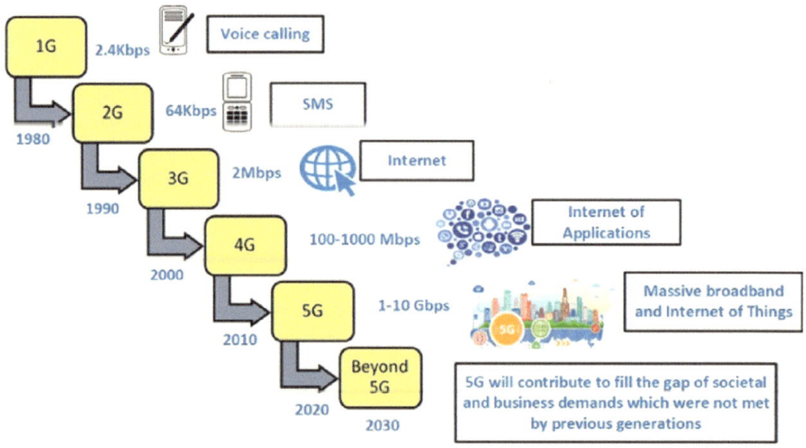

Figure 1.
The evolutionary growth of wireless technology in terms of data transmission capacity and applications.

across vast distances. As we venture into B5G and 6G, the role of antennas becomes even more pronounced, requiring designs that can cater to the unique demands of these advanced communication systems [2–8]. The journey toward B5G and 6G is not without challenges. The need for higher data rates, ultra-low latency, and unprecedented device connectivity introduces complexities that demand innovative solutions. Millimeter-wave frequencies and the exploration of terahertz domains present new frontiers, each with its own set of hurdles that antenna design must overcome [2, 9–12]. This chapter seeks to unravel these challenges and illuminate the path toward antennas that can navigate the uncharted territories of advanced wireless communication.

As we push the boundaries of connectivity, the electromagnetic spectrum becomes both a canvas and a constraint. This chapter delves into the strategic utilization of spectrum resources, exploring the nuances of millimeter-wave frequencies and delving into the potential of terahertz bands. Concepts like beamforming, massive MIMO, and intelligent antenna arrays come to the force, offering a glimpse into the technologies that will define the antennas of B5G and 6G [13–17].

As we embark on this journey through the realms of B5G and the anticipated era of 6G, this chapter beckons researchers, engineers, and enthusiasts to join the exploration of antenna evolution. By providing a comprehensive overview of the state-of-the-art in antenna design, this chapter aims to empower those at the forefront of technological innovation, equipping them with the knowledge and insights needed to shape the wireless communication landscape of the future.

Join us as we unravel the mysteries of antenna evolution and step into the transformative potential that awaits in the realm of Beyond 5G and 6G applications.

2. Antenna evolution in B5G and 6G

Having established the groundwork, let us delve into the evolving landscape of antenna design tailored for B5G and 6G applications. Antennas in these advanced networks are not just communication conduits; they are catalysts for innovation. The spectrum of possibilities broadens as we explore unconventional designs, including compact microstrip antennas, metamaterial-based antennas, reconfigurable antennas, and intelligent adaptive arrays. These designs aim to address the unique challenges posed by the increasing the demand for higher data rates, lower latency, and massive device connectivity.

3. Microstrip antennas

Microstrip antennas emerge as key contenders in this landscape, offering a versatile and scalable solution for next-generation communication networks. Microstrip antennas are planar antennas that find widespread applications in modern communication systems. Their inherent advantages, including low profile, ease of fabrication, and compatibility with integrated circuits, make them particularly appealing for B5G and 6G scenarios. Microstrip antennas are inherently compact, making them suitable for applications where size constraints are critical, such as in small and wearable devices. Microstrip antennas can be seamlessly integrated with integrated circuits, enabling the development of highly integrated and miniaturized communication modules for B5G and 6G devices. Microstrip antennas can be designed to operate across multiple frequency bands, facilitating the diverse spectrum requirements of advanced communication systems [18–21]. It is anticipated that two frequency bandwidths are allocated for B5G and 6G applications. The lower spectrum is estimated to be from 3 GHz to 20 GHz, and the upper spectrum is estimated to be from 0.1 THz to 10 THz to satisfy the ever-increasing demand of data transmission capacity required for emerging technologies such as artificial intelligence (AI), virtual reality, Internet of Things, biomedical sensors, and microwave imaging. Recently, many techniques are introduced to enhance the bandwidth of the microstrip antennas and achieve super-wideband antennas covering frequencies from 2.5 GHz up to more than 20 GHz which is suitable for the lower frequency band of B5G and 6G applications. **Figure 2** shows the geometry of a microstrip slot antenna, applying a pair of T-shaped slits in the ground plane to enhance the antenna's bandwidth and extend it to 23 GHz [22]. In [23], a pair of semi-circle slots in the ground plane along with a truncated radiation patch are applied to highly enhance the bandwidth of the antenna to more than 23 GHz as shown in **Figure 3**. On the other hand, because of very wide bandwidth of the above-mentioned antennas, interferences can occur with other coexisting narrow band systems operating within 3–20 GHz such

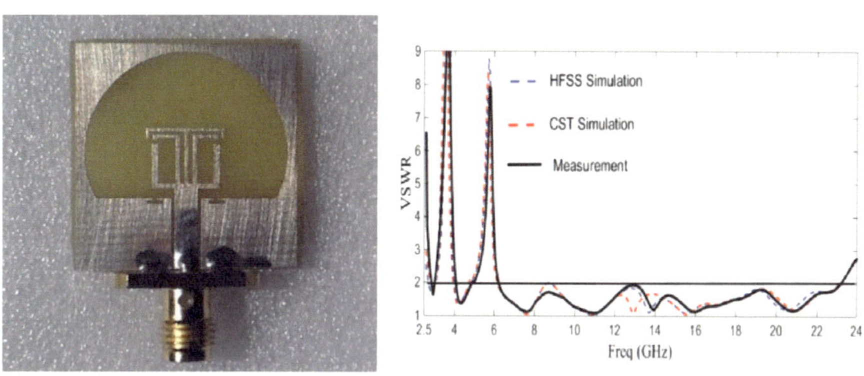

(a) (b)

Figure 2.
(a) Fabricated super wideband antenna presented in [22], and (b) VSWR of the antenna.

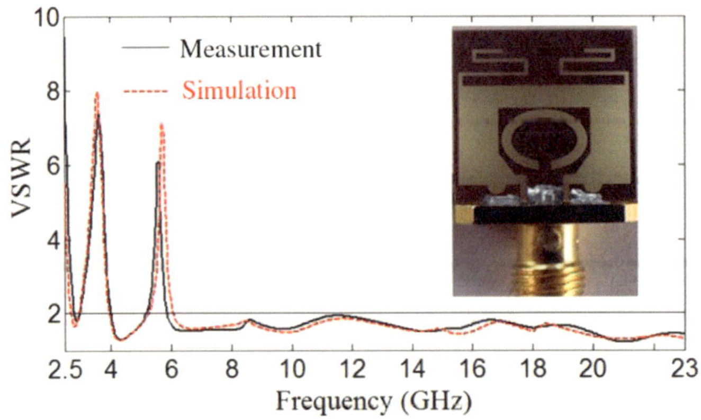

Figure 3.
Super wideband antenna proposed in [23] and its VSWR.

as WiMAX (3.3–3.7 GHz) and WLAN (5–6 GHz) systems. To avoid these interferences, various filtering structures are incorporated inside these antennas' geometry. As displayed in **Figure 2**, in [22], two inverse L-shaped stubs and a central T-shaped stub are connected to the fork-shaped radiation patch in order to achieve dual band-notched properties. In [23], two S-shaped slits are cut in the ground plane, and a circle-like ring slot is etched in the radiation patch to obtain band-rejected properties. It can be seen from **Figure 2(b)** and **Figure 3** that two peaks in the VSWR diagrams can occur at WiMAX and WLAN spectra due to the presence of these filtering structures. As mentioned in [23], the wide bandwidth characteristics of the above-mentioned antennas can be applied in impulsive systems for biomedical and healthcare applications where ultra-short pulses are required to detect cancer cells and tissues in the breast and brain. In [24], the effect of a super wideband antenna with triple band-notched characteristics on different types of pulses including modulated Gaussian (MG) and square-root-raised cosine (SRRC) pulses in different scenarios is investigated. The antenna has a bandwidth of 2.5–20 GHz and provides triple band-notched properties (see **Figure 4**). To assess the capability of the antenna for sending and receiving short pulses without distortions, the received pulses and their frequency responses are simulated and measured as shown in **Figure 5**. The system fidelity factor which is a measure of similarity between two signals should be more than 50% to gain a reliable wireless communication. The antenna and

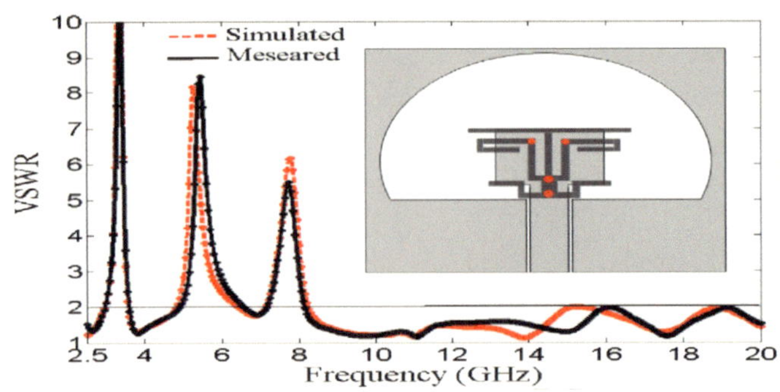

Figure 4.
Geometry and VSWR of the antenna introduced in [24].

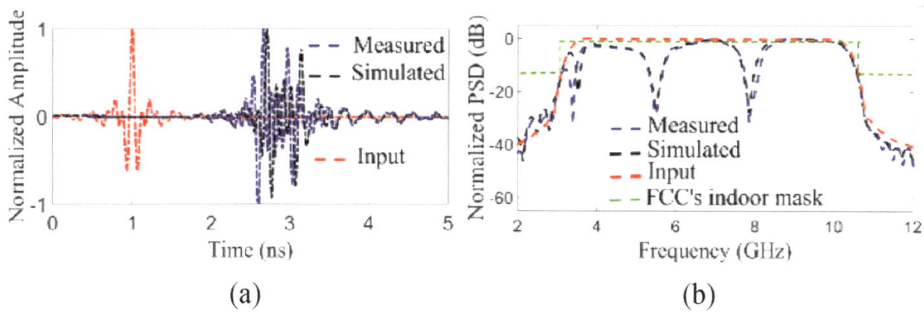

Figure 5.
The effect of the antenna in [24] on SRRC pulse: (a) time domain and (b) frequency domain in the face-to-face scenario.

filtering structures applied into the antenna geometry can cause distortions and ringing effects on transmitting and receiving pulses.

In [24], the system fidelity factor values are more than 80% for SRRC pulses in different scenarios indicating the suitability of this antenna for B5G and 6G impulsive radio applications. Additionally, some microstrip antennas are designed to cover the terahertz (THz) spectrum [25–30]. **Figure 6** demonstrates a recently designed microstrip antenna operating at a bandwidth of 345–375 GHz [25]. The antenna is designed on an A6 M Ferro LTCC substrate with ε_r = 5.99, thickness of 75 μm, loss tangent of 0.002, and the overall size of 600 × 600 μm^2. A stepped microstrip feed-line and a rectangular radiation patch is employed to achieve acceptable gain and bandwidth in the THz spectrum as shown in **Figure 6(b)** and **(c)**. In [26], a Z-shaped patch antenna is realized on Rogers 6006 substrate with ε_r = 6.15, thickness of 127 μm, and loss tangent of 0.0027 providing dual-band and left-handed circularly polarized (LHCP) characteristics as shown in **Figure 7**.

The antenna is covering 0.19–0.24 THz and 0.57–0.59 THz bands, with a maximum gain of 5.8 dBi. The LHCP properties can be achieved at 45° in boresight.

Despite their advantages, microstrip antennas face challenges related to radiation gain and directivity to overcome the severe path losses in B5G and 6G propagation environments. Researchers and engineers are actively addressing these challenges through innovative design techniques, such as metamaterial integration, advanced feeding techniques, and novel substrate materials.

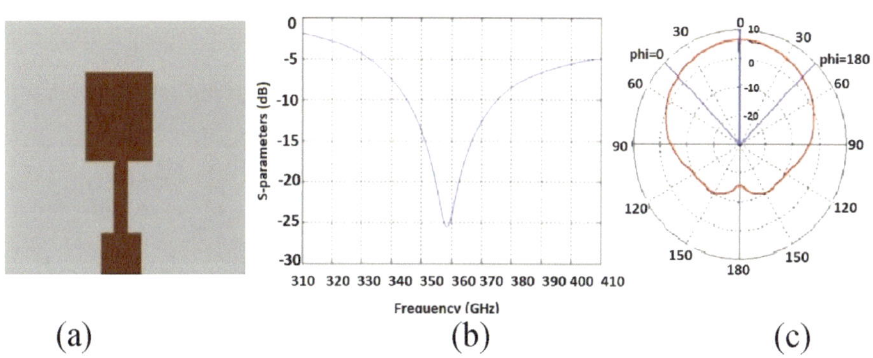

Figure 6.
(a) Geometry of the rectangular THz microstrip antenna. (b) Return loss and (c) radiation pattern.

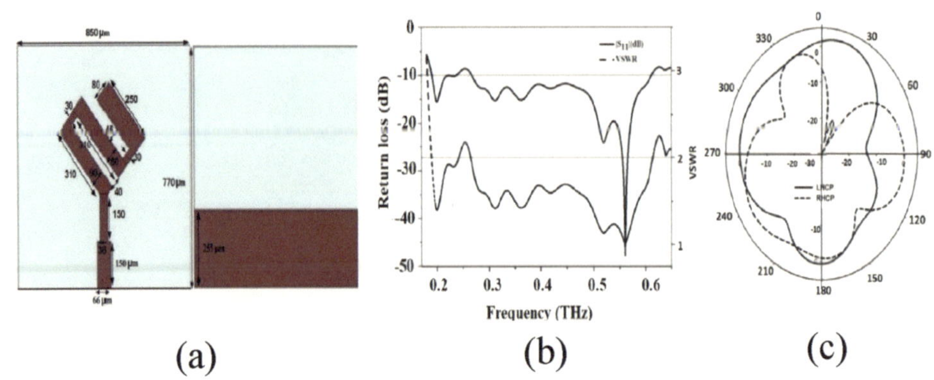

Figure 7.
(a) Geometry of Z-shaped patch THz microstrip antenna. (b) Return loss and (c) radiation pattern.

3.1 Metamaterial array antennas

As indicated in the previous section to solve the path loss and atmospheric absorption in B5G and 6G frequency regimes, one way is to employ array antennas providing high directivity and radiation gain. To more clarify how the array works please consider **Figure 8**. In **Figure 8(a)**, a single isotropic antenna and its radiation at half of the pattern is depicted. The antenna radiates out a sine wave, where the light curves are where the signal has high power and the dark lines are where the signal has low power. This antenna radiates equally in all directions; therefore, it has no directivity. In **Figure 8(b)**, we have an array configuration including two elements that are placed half a wavelength away and are also radiating out the exact same sine wave. Along the horizontal axis, it can be seen that the two waves are exactly out of phase with each other since the light part of one overlaps the dark part of the other. This means that when one signal is high, the other is low, and when those two signals are combined, they cancel out each other and very little energy in these directions remains. However, if we look in the vertical direction, it can be seen that the two signals are almost in phase with each other. This means both signals are low or high at the same time, and they constructively add in this direction. Recently, many array antenna configurations for mm wave and THz applications are presented [26–32].

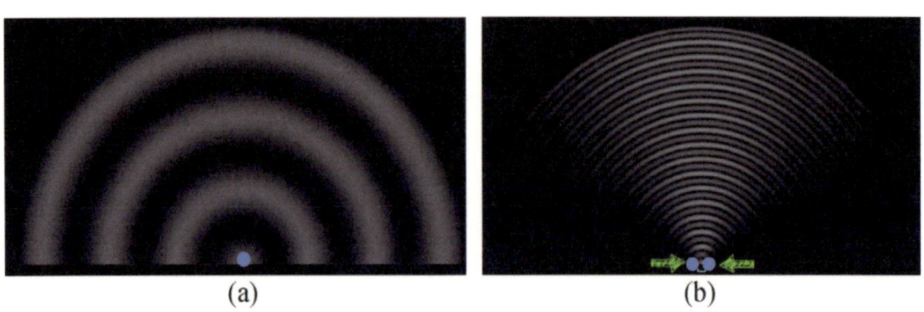

Figure 8.
(a) Radiation of a single element omnidirectional antenna and (b) radiation of a two-element array antenna.

In [26], a 2 × 32 antenna array is designed and fabricated to operate in 37.5 GHz. The antenna is realized on the Rogers 5880 substrate with a relative permittivity of 2.2 and loss-tangent of 0.0009.

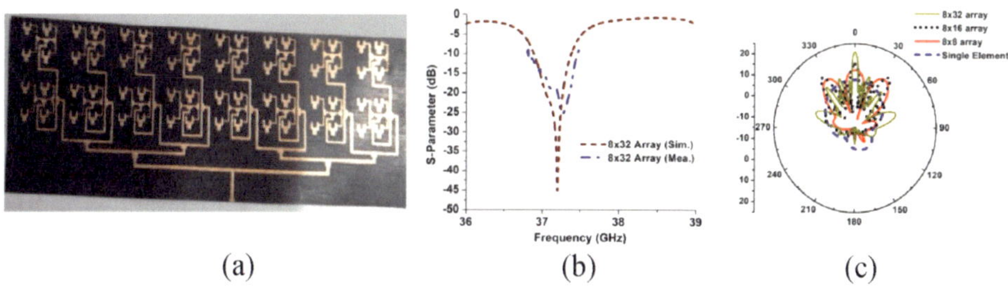

Figure 9.
(a) Geometry of 64-element array antenna. (b) Return loss and (c) radiation pattern in boresight [27].

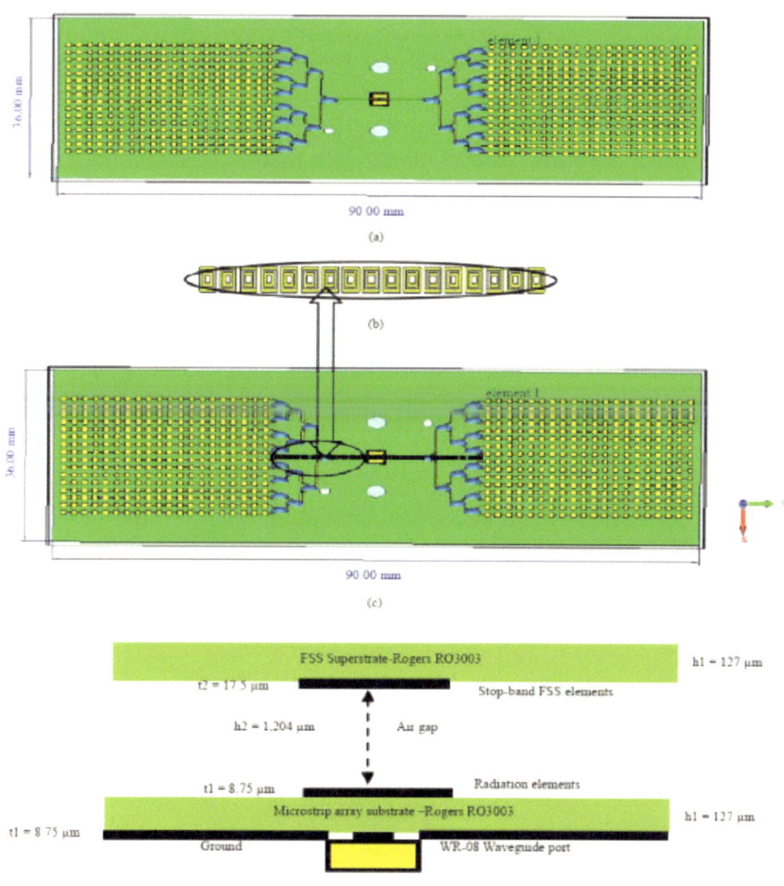

Figure 10.
(a) Geometry of an array antenna in [29], (b) metamaterial surface, (c) array antenna with metamaterial superstrate over microstrip feed-line, and (d) cross-sectional view of the antenna with metamaterial layer poisoned over it.

The single element employed to build the array construction is a fork-shaped patch microstrip antenna with a gain of 7.6 dBi, while the full array antenna with 64 elements exhibits a gain of 21.2 dBi. **Figure 9(a)** shows the fabricated array antenna, and **Figure 9(b)** and **(c)** demonstrate the return loss and radiation patterns of the array antenna with different numbers of elements.

Metamaterial antennas enable the creation of compact and lightweight designs, facilitating miniaturization of devices [28]. This is particularly advantageous in the context of B5G and 6G technologies, where there is a growing demand for smaller and more directive communication devices.

Metamaterials offer unique electromagnetic properties that can be tailored to enhance the antenna performance. This includes improved radiation gain, directivity, and bandwidth, addressing the demand for higher data rates and spatial coverage in advanced wireless networks. Metamaterial antennas can be designed to operate across multiple frequency bands, supporting the diverse spectrum requirements of B5G and 6G technologies. This flexibility is crucial as these networks integrate a wide range of frequency bands to meet the demands of various applications.

Metamaterials can be engineered to adapt to changing environmental conditions and communication scenarios. This adaptability is beneficial in B5G and 6G networks, where dynamic spectrum sharing, massive connectivity, and diverse communication modes are expected to be prevalent. However designing metamaterial antennas have several challenges including complex design and fabrication, sensitivity to manufacturing tolerances, and frequency band limitations. Usually, an array of unit cells which produce un-natural constitutive material characteristics (negative primitively and permeability) is employed to improve the properties of the antenna. **Figure 10** demonstrates that a layer of rectangular split ring resonators as metamaterial unit-cells is placed over the antenna to enhance the gain of the antenna [29]. The antenna consists of a pair of 1 × 16 Wilkinson power dividers and a pair of 16 × 20 radiation patches realized on a Rogers 4003 substrate to operate at the THz spectrum (115 GHz). **Figure 11** shows the 3-dimensional radiation patterns of the antenna with

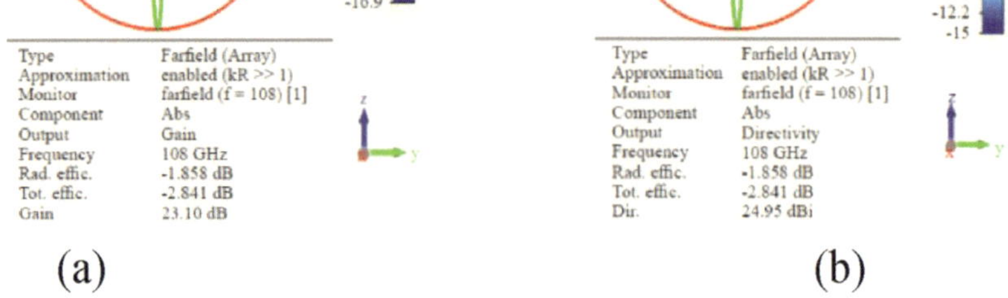

Figure 11.
3-dimensional radiation pattern of the antenna (a) without metamaterial superstrate and (b) with metamaterial superstrate [29].

DOI: http://dx.doi.org/10.5772/intechopen.1005483

and without metamaterial superstrate at 115 GHz. As we can see from the figure, the radiation gain of the antenna is improved about 1 dB in the presence of the metamaterial layer.

4. Beamforming and beam steering

In addition to the fact that the antennas must have high gains, they must have the ability to steer the main beam because 5G and 6G cellular networks have many mobile users and it requires a very reliable link as indicated in **Figure 12(a)**. Also, there is the possibility of blocking line of sight connection in crowded and dense environments. The capability of the rotating beam provides the connection from other routes as shown in **Figure 12(b)**.

4.1 Pattern reconfigurable array antennas

Pattern reconfigurable array antennas are a promising technology for 6G wireless communication systems, offering the ability to adapt their radiation patterns dynamically to meet the requirements of different communication scenarios. These antennas play a crucial role in enhancing the performance, reliability, and efficiency of communication systems in the 6G era. These antennas enable dynamic beamforming, allowing the antenna system to adjust the direction of its radiation pattern in real-time. This capability is vital for 6G networks that will require efficient beam steering to support diverse communication scenarios. Pattern reconfigurable antennas can be designed to adapt to different frequency bands, ensuring compatibility with the diverse spectrum requirements of 6G.

In [30], a conformal reconfigurable array antenna for mm wave applications is reported. The overall antenna geometry in both cylindrical and planer configurations as well as a 2 × 1 unit-cell is depicted in **Figure 13**. The antenna consisting of a 2 × 16 planar array is fabricated on a polyimide substrate with height of 0.27 mm which is conformal. It is noted that the planar array's beam steering covers less than 180°. However, such beam coverage seems to be sufficient, electronic devices are mainly

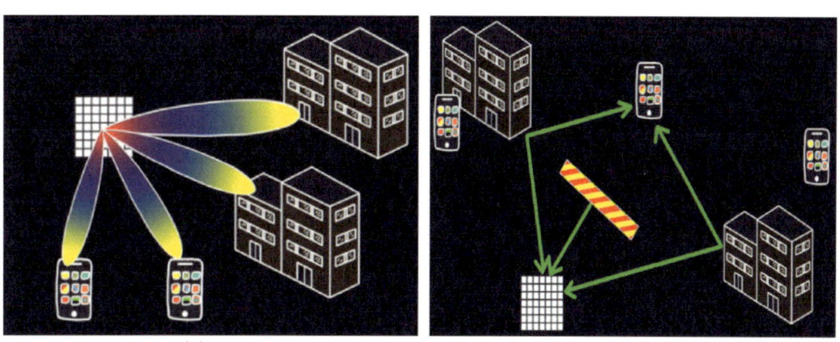

(a) (b)

Figure 12.
The advantages of beam steering in B5G and 6G networks: (a) multiuser capacity and (b) multipath channel capacity.

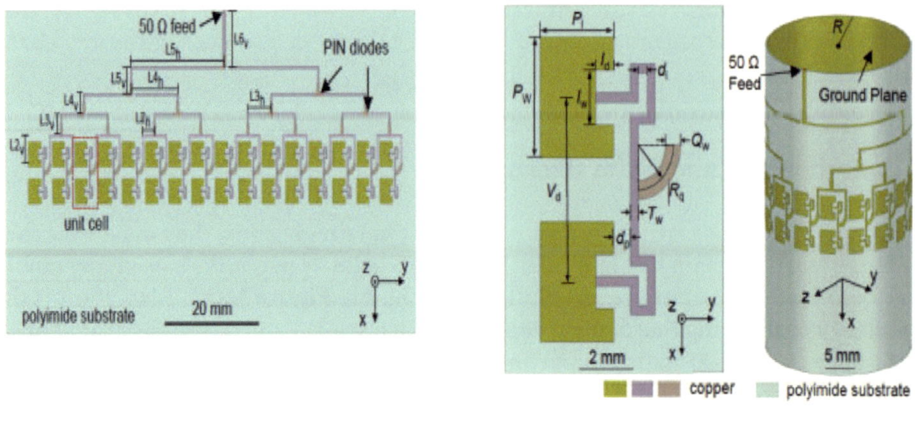

(a) (b) (c)

Figure 13.
Pattern reconfigurable array antenna (a) planar configuration, (b) a 2 × 1 unit-cell, and (c) cylindrecalconfiguration [30].

subjected to dynamic orientations by their users and can be situated in orientations where individuals unintentionally block signals. To address this problem and achieve a 360° coverage, the array is wrapped around a cylindrical structure with an outer polyimide substrate layer and an inner ground plane. The array is fed in parallel by using microstrip transmission lines. In each state, only 2 × 2 unit cells are fed. 14 PIN diodes are employed to switch between all eight states. **Figure 14** demonstrates the return loss and radiation patterns of the antenna for different states. Clearly, return loss curves are acceptable for all states at 28 GHz, and the radiation patterns dynamically cover 360°.

4.2 Phased array antennas

As we discussed earlier, it is evident that a really sharp beam can be formed with an array of antennas [31]. Generally, the sharper main lobe can be achieved by

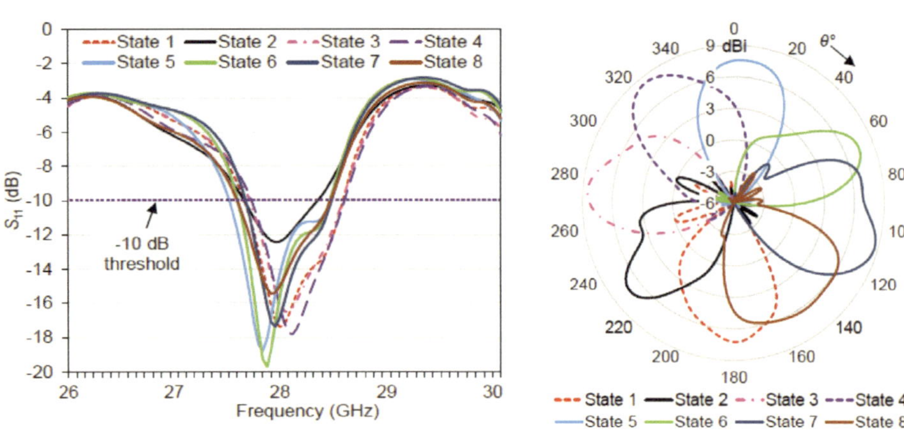

Figure 14.
(a) Return loss and (b) radiation pattern of antenna in [30] at 8 different states.

DOI: http://dx.doi.org/10.5772/intechopen.1005483

employing more elements in the array. In this section, it is worthy to know how the shape of the beam can actually be more controlled rather than just adjusting the position and orientation of the main beam. In phased array antennas, by phase

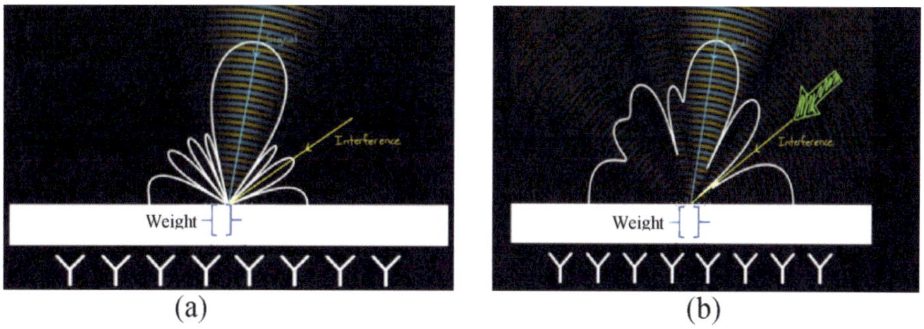

Figure 15.
Beamforming of the phased array antenna: (a) without interference mitigation and (b) with interference mitigation.

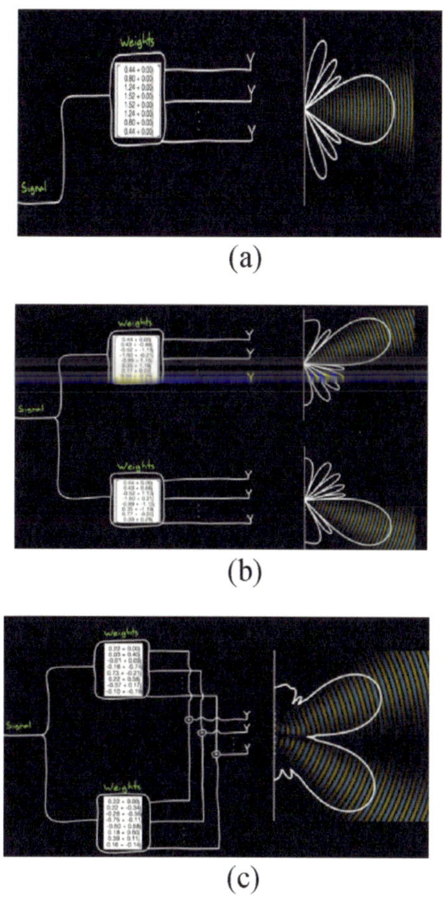

Figure 16.
(a) A single phased array antenna, (b) two separated phased array antennas, and (c) two merged phased array antennas.

shifting, the beam can be steered. Nevertheless, creating the main lobe and positioning it at the desired signal is not the only issue we care about in the phased array antenna. In this case, the quality of the signal, and particularly how much signal can be received compared to noise and interference, is important. And, the environments are full of noise and interference. Considering **Figure 15**, now, without regarding the source, potentially, we have a desired signal in some direction that we would like to maximize (shown in blue) and interference in others that we want to minimize (indicated by yellow). Ideally, the objective is to direct a focused beam toward the signal, aiming for low array gain in the interference direction. The goal is to achieve a satisfactory signal-to-noise-plus-interference ratio at the receiver. However, it is important to acknowledge the presence of side lobes, which, although having lower gain than the main lobe, can still be impactful if there is a strong interference signal from that specific direction. In a phased array antenna, we have the ability to regulate side lobe levels by adjusting the amplitude of each element. By modifying the phase of individual elements, we can steer the main lobe toward the desired signal direction. Simultaneously, fine-tuning the element amplitudes allows for the suppression of interferences while maintaining the constancy of the main beam. The weight vector illustrated in **Figure 15** serves as both a phase and amplitude tuner. **Figure 16** demonstrates that, by combining several weight vectors, we can attain multiple-beam properties, enabling the simultaneous coverage of more users and systems. In [32], a phased array antenna system with separated phase and amplitude electrical controllers is presented. The geometry of the antenna is shown in **Figure 17** including 16 subarrays, 16 beamforming receiving RF (BF-RFM) units, a control/power board, and a 16-way feed network. The 16-dipole antenna subarray was partitioned into two 8x1 subarrays, each linked to its respective BF-RFM. The

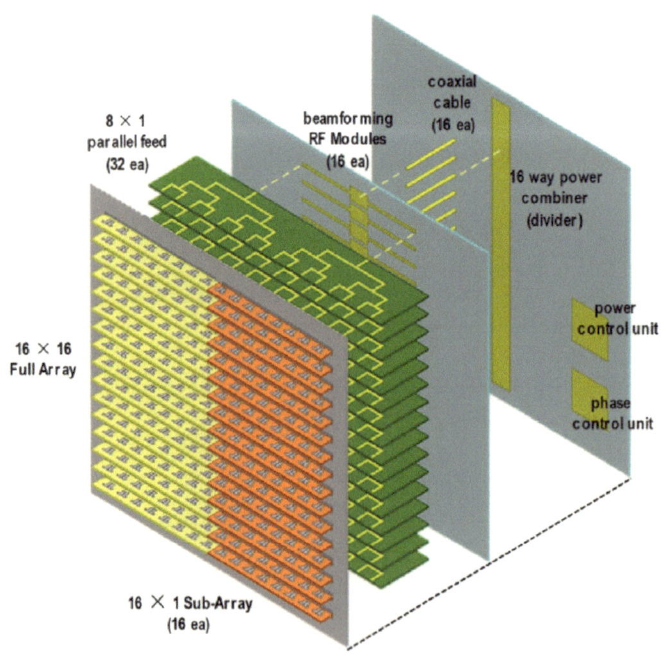

Figure 17.
A single phased array antenna, (b) two separated phased array antennas, and (c) two merged phased array antennas [32].

DOI: http://dx.doi.org/10.5772/intechopen.1005483

development of the phased-array antenna involved the individual design, fabrication, and measurement of each antenna subsystem. Low-loss and in-phase coaxial cables were utilized to connect the subarrays to their corresponding 16 BF-RFMs. Similarly, 16 coaxial cables were employed to establish connections between the BF-RFM output ports and the 16-way feed network.

The power/control board was specifically designed to supply power and control bits to the 16 BF-RFMs. As depicted in **Figure 18**, the BF-RFM comprised a two-way power combiner for joining two 8x1 arrays, three low-noise amplifiers (LNAs) to ensure low-noise performance and adequate system gain, a phase shifter, and a digital attenuator for modifying the phase and amplitude and of each radiating element. Utilizing low-loss and in-phase coaxial cables, the two input ports (Pin 1 and Pin 2) were merged with two 8 × 1 arrays, while the single output port (Pout) was linked to the 16-way feed network. **Figure 19** shows the radiation pattern of the antenna in the azimuth and elevation plane. It can be seen that the main beam and side-lobe levels can be easily controlled.

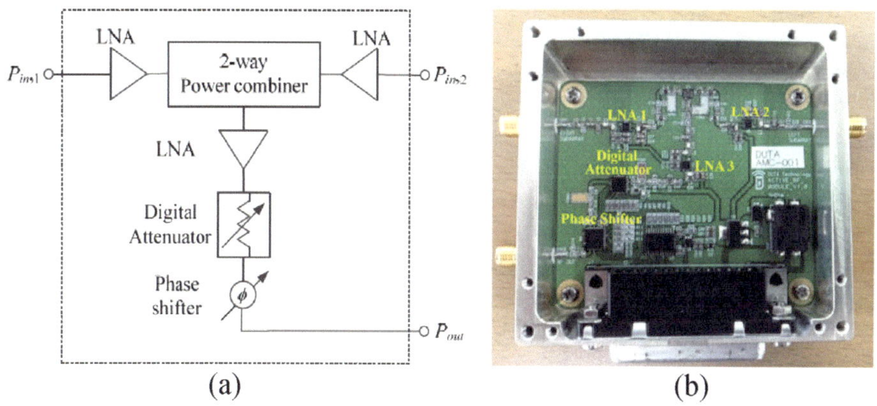

Figure 18.
(a) Schematic and (b) image of beamforming receiving RF [32].

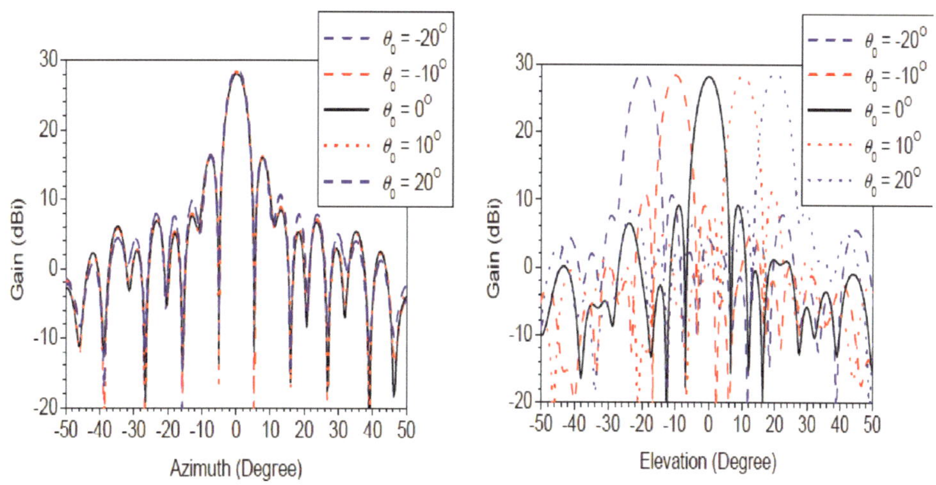

Figure 19.
(a) Schematic and (b) image of beamforming receiving RF [32].

4.3 Lens antennas

Lens antennas, a crucial component in advanced communication systems, play a pivotal role in B5G and 6G applications. These antennas leverage electromagnetic lenses to manipulate signals, enhancing communication capabilities in unprecedented ways. The design of lens antennas for B5G and 6G era necessitates careful consideration of frequency bands, beamforming requirements and seamless integration with evolving communication systems. Adapting to the unique demands of these networks is essential for optimal performance.

Exploring materials tailored for lens antennas and employing cutting-edge manufacturing methods becomes imperative. Advancements in material science and fabrication techniques contribute to achieving the precision required for these next-generation communication systems. Lens antennas contribute significantly to beam steering and beamforming in B5G and 6G networks. Their ability to focus signals directionally enhances the efficiency and reliability of communication, supporting the dynamic requirements of these advanced systems. Synergistic integration with metamaterials amplifies the capabilities of lens antennas in the future wireless networks.

Addressing challenges specific to lens antennas in B5G and 6G networks is paramount. Robust solutions, including advanced signal processing algorithms and adaptive control mechanisms, are key to overcoming obstacles and ensuring optimal antenna performance.

Various lens-based beamformers including Ruze, Rotman, and Luneburg structures are employed to design lens antennas. Ruze lenses demonstrate multiple authentic focal points, resulting in minimal phase aberrations across a broader scanning range. However, the constrained design of these lenses introduces limitations due to port discretization, particularly influenced by the technology of the transmission line. Seeking to address these constraints, a novel approach known as the continuous parallel-plate waveguide (PPW) beamformer was introduced [32]. This beamformer transforms the cylindrical wave, launched by a primary feed (e.g., sectoral horns in **Figure 20(a)**), as it propagates within the PPW section, converting it into a quasi-plane wave radiated into free space, and vice versa. The required delay correction for this wave transformation is achieved through a PPW lens with transversal ridge and cavity features (see **Figures 20(b)** and **19(c)**, labeled 1 and 2 for inner and outer contours). By employing a bifocal constrained lens approximation, this design is anticipated to exhibit two focal points, F1 and F2 (as shown in **Figure 20(a)**), primarily regulated by adjusting the ridge height (h_w in **Figure 20(c)**).

Rotman lenses have found widespread utility in wireless communication systems. However, the conventional version tends to be excessively large, particularly at lower operating frequencies, making integration with base station antennas in cellular communication systems challenging when compared to circuit-based beamforming techniques. To address this size constraint, an innovative approach was proposed in [34] involving the use of dielectric material to load the parallel plate region of the Rotman lens, aiming to reduce its overall dimensions. The authors detail in Ref. [34] the realization of a compact Rotman lens-fed antenna array designed for cellular wireless applications. This lens-fed antenna adopts a multi-layer configuration, consisting of two layers, resulting in a design with a low profile and a relatively small footprint. This compact form factor makes it particularly appealing for applications where size is a crucial factor. The prototype of this designed antenna is illustrated in **Figure 21**.

In the mentioned systems, inherent phase errors exist across the apertures for the majority of multiple beams due to constrained lens designs, which typically

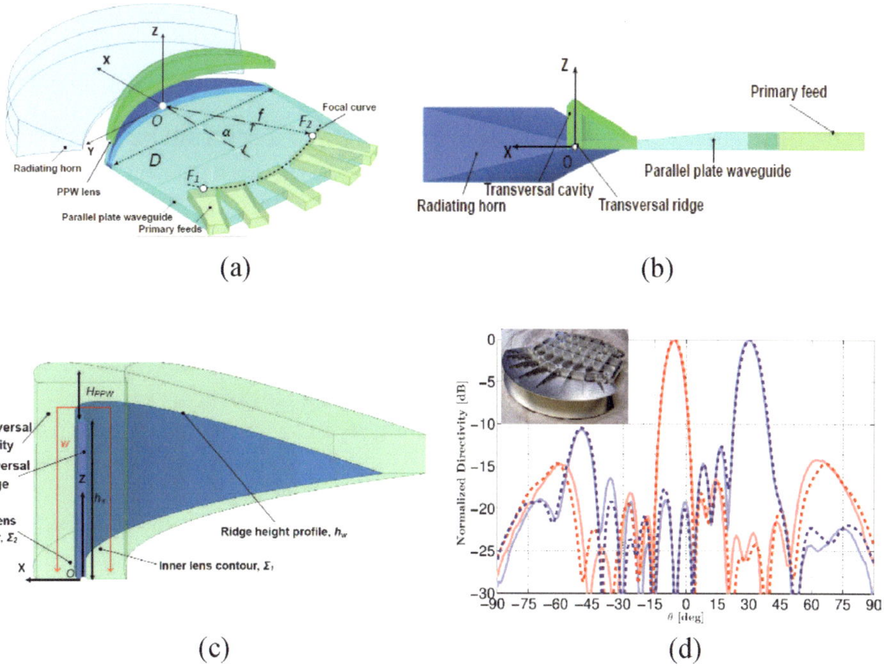

Figure 20.
PPW lens beamformer: (a) 3D view, (b) cross sectional view, (c) focus on the transversal cavity, and (d) simulated (dash line) and measured (solid line) radiation pattern of the manufactured PPW beamformer in the H-plane at fo = 10.7 GHz [33].

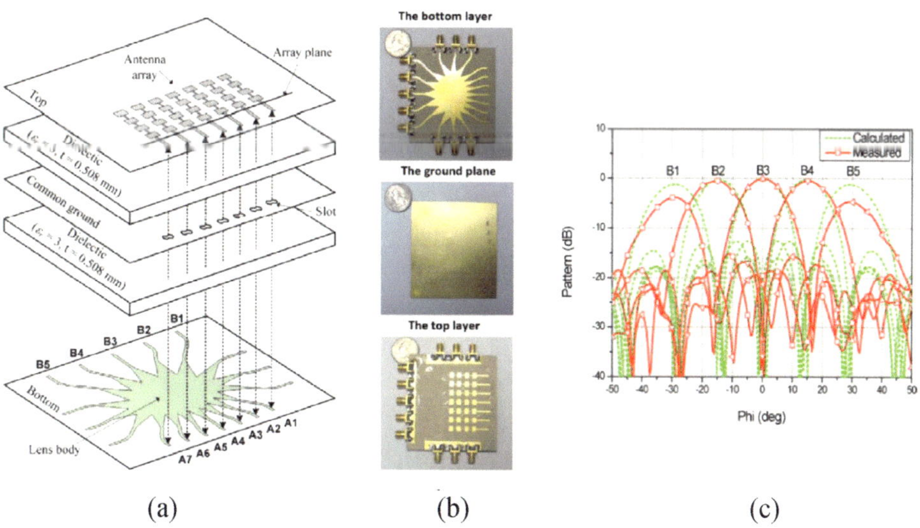

Figure 21.
Two-layer Rotman lens-fed antenna presented in [34]: (a) geometry, (b) fabricated prototype, and (c) simulated and measured radiation pattern.

offer only a limited number of true focal points, resulting in a scanning range typically not exceeding ±50 degrees. In contrast, Luneburg lenses and their derivatives emerge as highly desirable multi-beam lens antennas since, theoretically, they exhibit no phase errors for any pointing direction. Recent advancements in 3D printing techniques, known for high accuracy fabrication at a low cost, have garnered significant attention from researchers focusing on Luneburg lenses. An illustrative example is found in Ref. [35], where a novel 3D printed Luneburg lens, fed by a dual-polarized magnetoelectric (ME) dipole, was introduced. The design prototype successfully generated nine beams across the entire Ka-band, making it a promising candidate for multi-beam cellular applications. The structural details are depicted in **Figure 22**.

4.4 Massive MIMO arrays

Massive multiple input and multiple output (MIMO) array leverages the spatial dimension to serve multiple users concurrently by transmitting unique data streams to each user on the same time-frequency resource; in other words, it crates multiple channel communications [36, 37]. This spatial multiplexing capability significantly improves the spectral efficiency. The use of a large number of antennas and spatial processing techniques allows for efficient beamforming, which enhances the link quality and spectral efficiency of the communication system required for B5G and 6G era. Massive MIMO can provide improved signal quality, reduced interference, and better coverage, especially in challenging environments with obstacles or signal

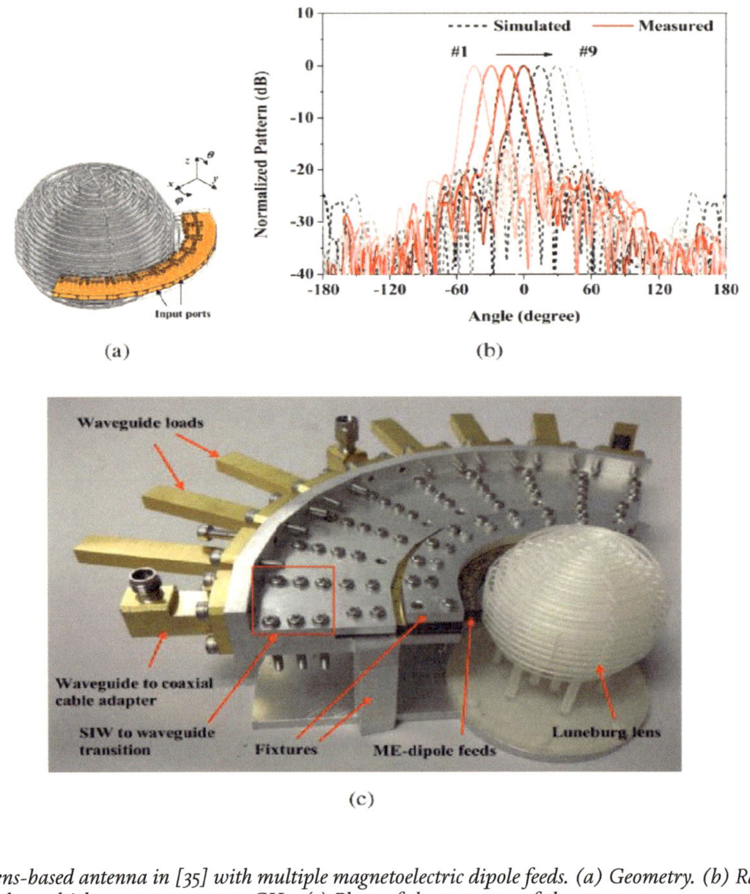

Figure 22.
Luneburg lens-based antenna in [35] with multiple magnetoelectric dipole feeds. (a) Geometry. (b) Radiation patterns of the multi-beam antenna at 32 GHz. (c) Photo of the prototype of the antenna.

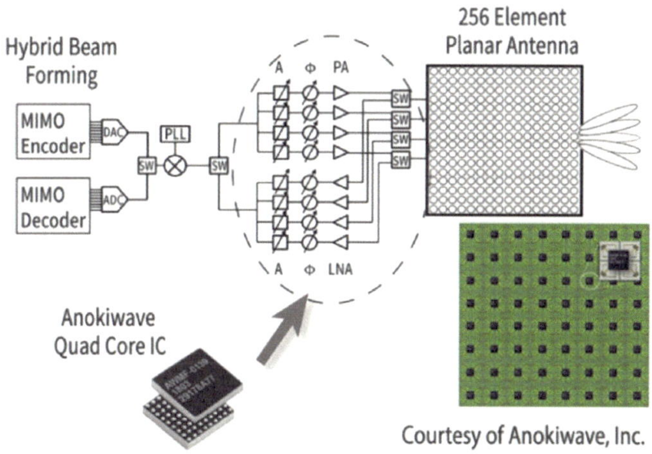

Figure 23.
An instance of massive MIMO arrays for cellular networks.

Ref	Type	Operating frequency	Technique
[24]	Single element	3 GHz–20 GHz	Parasitic element
[26]	Single element	0.19–0.24 THz	Z-shaped radiation patch
[27]	Array	37–37.5 GHz	Fork-shaped patch
[29]	Array	111–115 GHz	Metamaterial
[30]	Array	27.5–28.5 GHz	Reconfigurable
[32]	Array	2.2–2.6 GHz	Phased array
[33]	Single	10.7 GHz	Parallel-plate waveguide
[35]	Single	32 GHz	Luneburg lens
[36]	Massive MIMO	100 GHz	Constellation of 32 antennas

Table 1.
Characteristics of recently reported antennas for B5G and 6G.

blockage. Also, massive MIMO systems can achieve energy efficiency gains by exploiting spatial multiplexing and beamforming, leading to better utilization of resources and reduced interference. **Figure 23** illustrates a massive MIMO array with 256 elements. The Anokiwave AWMF-0139 IC, which houses four antenna modules, serves as a transceiver module. A total of 64 of these ICs are connected to the 256 elements, processing and generating various signals while controlling different beams to cover multiple users in the 5G and 6G eras.

Table 1 summarizes types and characteristics of the antennas discussed in this chapter for B5G and 6 G applications.

5. Conclusion

In conclusion, this chapter explored the fundamental concepts of various antenna systems critical for beyond 5G (B5G) and anticipated 6G technologies. We delved into microstrip antennas, metamaterials, reconfigurable antennas, phased array, and lens antennas, highlighting their contributions to achieving the ultra-high data rates, ultra-low latency, and massive device connectivity envisioned for B5G and 6G networks. These advancements in antenna design are not only crucial for efficient communication but will also pave the way for the development of novel B5G and 6G applications that will redefine how we connect, interact, and experience the digital world.

Author details

Seyed Ramin Emadian
Independent Researcher, Tehran, Iran

*Address all correspondence to: ramin.emadian@gmail.com

References

[1] Rikkinen K, Kyosti P, Leinonen ME, Berg M, Parssinen A. THz radio communication: Link budget analysis toward 6G. IEEE Communications Magazine. 2020;**58**(11):22-27

[2] Akyildiz IF, Jornet JM, Han C. Terahertz band: Next frontier for wireless communications. Physical Communication. 2014;**12**:16-32

[3] Shahen SAFM, Al-Bu Mohammed DJ, Pathan AK, Rabie K. An effective MIMO antenna design for 5G and beyond to support industry 5, communication, networking and broadcast technologies. TechRxiv. 2023

[4] Adamec B, Machaj J, Brida P. A novel wideband splitter for a four-element antenna array. Applied Sciences. 2024;**14**(4):1593

[5] Zhang J, Zhang Q, Qin W, Yang W, Chen J. Compact and broadband substrate integrated dielectric resonator antenna suitable for 5G millimeter-wave communications. IEEE Open Journal of Antennas and Propagation. 2023;**4**:982-989

[6] Zeng Q, Chen Z, He M, Wang S, Liu X, Xu H. Design of a Ka-band Heterogeneous integrated T/R module of phased array antenna. Electronics. 2024;**13**(1):204

[7] Abdelghany M, Ibrahim A, Mohamed H, Tammam E. Compact sub-6 GHz four-element flexible antenna for 5G applications. Electronics. 2024;**13**(3):537

[8] Yin Y, Xia C, Liu S, Zhang Z, Chen C, Wang G, et al. Ultrathin antenna-in-package based on TMV-embedded FOWLP for 5G mm-wave applications. Electronics. 2024;**13**(5):839

[9] Chen Z, Ma X, Zhang B, Zhang Y, Niu Z, Kuang N, et al. A survey on terahertz communications. China Communications. 2019;**16**(2):1-35

[10] Watanabe AO, Ali M, Sayeed SY, Tummala RR, Pulugurtha MR. A review of 5G front-end systems package integration. IEEE Transactions on Components, Packaging, and Manufacturing Technology. 2020;**11**(1):118-133

[11] Wells J. Faster than fiber: The future of multi-G/s wireless. IEEE Microwave Magazine. 2009;**10**(3):104-112

[12] Nayak S, Patgiri R. 6g communication: Envisioning the key issues and challenges. arXiv. 2020

[13] Han C, Wu Y, Chen Z, Wang X. Terahertz communications (TeraCom): Challenges and impact on 6G wireless systems. arXiv. 2019

[14] David K, Berndt H. 6G vision and requirements: Is there any need for beyond 5G? IEEE Vehicular Technology Magazine. 2018;**13**(3):72-80

[15] Lee WC. Mobile Communications Design Fundamentals. John Wiley & Sons; 2010

[16] Rappaport TS. Wireless Communications: Principles and Practice. New Jersey: Prentice hall PTR; 1996

[17] Jagannatham AK. Principles of Modern Wireless Communication Systems. McGraw-Hill Education; 2015

[18] Ezhilarasan E, Dinakaran M. A review on mobile technologies: 3G, 4G and 5G. In: 2017 Second International

Conference on Recent Trends and Challenges in Computational Models (ICRTCCM). IEEE; 2017. pp. 369-373

[19] Dahlman E, Parkvall S, Skold J. 4G, LTE-Advanced Pro and the Road to 5G. Academic Press; 2016

[20] Lee J, Han JK, Zhang J. MIMO technologies in 3GPP LTE and LTE-advanced. EURASIP Journal on Wireless Communications and Networking. 2009;**2009**:1-10

[21] Malviya L, Panigrahi RK, Kartikeyan MV. Four element planar MIMO antenna design for long-term evolution operation. IETE Journal of Research. 2018;**64**(3):367-373

[22] Emadian SR, Ahmadi-shokouh J. Modified ground circle like slot antenna with dual band-notched characteristics for super UWB applications. Applied Computational Electromagnetics Society (ACES) Journal. 2015;**30**:436-443

[23] Emadian SR, Ahmadi-Shokouh J. Very small dual band notched rectangular slot antenna with enhanced impedance bandwidth. IEEE Transactions on Antennas and Propagation. 2015;**63**(10):4529-4534

[24] Emadian SR, Ahmadi-Shokouh J, Ghobadi C, Nourinia J. Study on frequency and impulse response of novel triple band notched UWB antenna in indoor environments. International Journal of Electronics and Communications. 2018;**96**:93-106

[25] El-Nawawy M, Allam AM, Korzec D. The design of a 0.35 THz microstrip patch antenna on LTCC substrate. Electrical and Electronics Engineering. 2011;**1**(1):1-4

[26] Ullah S, Ruan C, Haq TU. Z-shaped dual band circular polarized microstrip antenna for THz communication. In: 2019 44th International Conference on Infrared, Millimeter, and Terahertz Waves (IRMMW-THz). IEEE; 2019. pp. 1-2

[27] Sehrai D, Khan J, Abdullah M, Asif M, Alibakhshikenari M, Virdee B, et al. Design of high gain base station antenna array for mm-wave cellular communication systems. Scientific Reports. 2023;**13**:4907

[28] Sabban A. Novel meta-fractal wearable sensors and antennas for medical, communication, 5G, and IoT applications. Fractal and Fractional. 2024;**8**(2):100

[29] Nissanov U, Singh G. 6G Broadband and High Directive Microstrip Antenna with SIW and FSSs. Engineering Science & Technology; 2023

[30] Stroh A, Sigda M, Carbone K, Baun D, Tajin MAS, Bshara O, et al. A pattern reconfigurable conformal mmWave antenna for 5G applications. In: IEEE International Symposium on Antennas and Propagation and USNC-URSI Radio Science Meeting (APS/URSI). IEEE; 2021

[31] Li M, Chen S, Liu Y, Guo Y. Wide-angle beam scanning phased array antennas: A review. IEEE Open Journal of Antennas and Propagation. 2023;**4**:695-712

[32] Lee DH, Seo JW, Lee MS, Chung D, Lee D, Bang JH, Satriyotomo B, Pyo S. An S-band-receiving phased-array antenna with a phase-deviation-minimized calibration method for LEO satellite ground station applications Electronics. 2022;**11**(23):3847

[33] Doucet F, Fonseca NJG, Girard E, Legay H, Sauleau R. Analytical model and study of continuous parallel

plate waveguide lens-like multiple beam antennas. IEEE Transactions on Antennas and Propagations. 2018;**66**(9):4426-4436

[34] Lee W, Kim J, Yoon YJ. Compact two-layer Rotman lens-fed microstrip antenna array at 24 GHz. IEEE Transactions on Antennas and Propagations. 2011;**59**(2):460-466

[35] Li Y, Ge L, Chen M, Zhang Z, Li Z, Wang J. Multi-beam 3D printed Luneburg lens fed by magneto-electric dipole antennas for millimeter-wave MIMO applications. IEEE Transactions on Antennas and Propagations. 2018;**67**(5):2923-2933

[36] Ting P, Yu S, Huang Z, Wei C, Chi S, Lin C. Fronthaul optical links using sub-Nyquist sampling rate ADC for B5G/6G sub-THz Ma-MIMO beamforming. IEEE Access. 2023;**10**:236-243

[37] Elijah O, Abdul Rahim S, New W, Leow C, Cuman K, Geok T. Intelligent massive MIMO systems for beyond 5G networks: An overview and future trends. IEEE Access. 2022;**10**:102532-102563

Chapter 7

Mutual Coupling in the Antennas

Mohammadreza Rohaninezhad, Meysam Jalali Asdabadi, Changiz Ghobadi and Javad Nourinia

Abstract

Mutual coupling refers to the intricate interaction among discrete antenna elements within an array, often causing interference and leading to the deterioration of critical performance parameters. This degradation includes reduced gain, increased side lobe levels, and distorted radiation patterns, significantly impacting the overall system's capabilities and coverage. With the increasing demand for higher data rates and capacity in next-generation wireless communication systems, such as 5G and the emerging 6G, array antennas are essential for enabling beamforming and spatial multiplexing. However, mutual coupling among antenna elements presents a substantial challenge, affecting the efficiency and effectiveness of these systems. This chapter furnishes a comprehensive exposition on the multifaceted challenges posed by mutual coupling in the domain of array antennas. Furthermore, it undertakes a systematic examination of diverse techniques and strategies meticulously devised to ameliorate the adverse ramifications. In addition, this work undertakes an exhaustive analysis to discern the far-reaching influence of mutual coupling on critical performance metrics, encompassing radiation patterns, efficiency, and channel capacity, all in the context of the dynamic landscape of next-generation wireless communication systems epitomized by 5G and the forthcoming 6G.

Keywords: mutual coupling, array antenna, graphite, defected ground structure (DGS), electromagnetic bandgap (EBG), bandgap

1. Introduction

Mutual coupling in the context of antennas presents a substantial and pressing concern within the realm of 5G and 6G wireless communication systems. In light of the profound aspirations of these forthcoming networks, aimed at provisioning elevated data rates, augmented capacity, and enhanced connectivity, the performance of antennas assumes paramount significance. Mutual coupling specifically denotes the interaction that transpires between adjacent antennas arranged within an array configuration, invariably precipitating unwelcome repercussions. These repercussions encompass a distortion of radiation patterns, a diminution in antenna efficiency, and a compromise in channel capacity. In the specific framework of 5G and 6G applications, where the deployment of multiple antennas is indispensable for the realization of advanced techniques such as beamforming and spatial multiplexing, the adversarial influence of mutual coupling looms large over the collective system performance. Notably, it engenders interference among antenna elements, thereby inducing perturbations in the desired directionality of radiation patterns. This, in turn, leads to a dissipation of signal strength, an augmentation of interference, and a curtailment of the extant coverage. Moreover, mutual coupling's impact reverberates within the realm of antenna efficiency by virtue of the alterations it imposes upon impedance characteristics. The ramifications include power dissipation, a curtailment in signal quality, and an overarching diminution in system efficiency. In addition to these facets, the specter of mutual coupling extends its reach into the arena of channel capacity, effectively introducing correlations among antenna elements. These correlations curtail the system's ability to concurrently transmit and receive independent data streams. In response to these formidable challenges, rigorous research and development initiatives have been spearheaded to ameliorate the effects of mutual coupling within the domain of 5G and 6G antenna systems. Diverse methodologies, including decoupling structures, optimization of antenna element spacing, and the employment of sophisticated signal processing algorithms, are being meticulously examined and advanced. These endeavors collectively strive to abate mutual coupling's influence and bolster the performance metrics of the forthcoming generation of wireless communication systems.

2. Unraveling the enigma of wireless interference

In the domain of wireless communication, where rapid data transmission and uninterrupted connectivity have become essential, optimizing antenna performance assumes a central role. Among the factors impacting antenna performance, mutual coupling stands as a significant and intricate phenomenon profoundly influencing the quality of wireless signals. A comprehensive grasp of mutual coupling intricacies becomes parab mount for the effective design of radio frequency (RF) systems, particularly within the context of advanced technologies such as 5G and the forthcoming 6G. Mutual coupling, aptly named, denotes the interaction between two or more closely situated antennas, where one antenna's performance is affected by the presence of its neighboring antennas. This interaction engenders a spectrum of challenges, including reductions in antenna gain, increased side lobes, distorted radiation patterns, and compromised link quality. Thus, understanding mutual coupling intricacies proves pivotal in ensuring the reliability of wireless communication. Notably, array antenna configurations, widely employed in wireless systems, exhibit heightened susceptibility to mutual coupling. Array antennas offer enhanced directivity and gain through the manipulation of constructive and destructive interference between individual radiating elements. However, their close proximity can induce mutual coupling, leading to undesired consequences. In the context of 5G, where array antenna technology forms the cornerstone of massive multiple-input multiple-output (MIMO) systems, the understanding of mutual coupling takes on heightened importance. To maximize system capacity and spectral efficiency, 5G systems leverage the simultaneous transmission and reception of multiple data streams via a multitude of antennas. Nevertheless, mutual coupling can distort signals, precipitating interference and compromising overall system performance. As the research endeavors for the next-generation wireless technology, 6G, continue to evolve, the issue of mutual coupling assumes even greater prominence. With the aspiration of achieving higher data rates, ultra-low-latency communication, and pervasive connectivity, 6G seeks to redefine the boundaries of wireless communication. However, to realize these ambitious objectives, the challenges posed by mutual coupling must be systematically addressed.

In wireless communication, optimizing antenna performance is crucial for rapid data transmission and uninterrupted connectivity. Mutual coupling, the interaction between closely situated antennas, significantly influences wireless signal quality, especially in advanced technologies like 5G and the upcoming 6G. Array antennas, common in wireless systems, are particularly susceptible to mutual coupling, affecting their directivity and gain. Managing mutual coupling requires a multifaceted approach, including decoupling techniques (passive and active), antenna design optimization, and simulation tools. By addressing mutual coupling, engineers can enhance wireless communication reliability and performance, which is crucial for advancing technologies like 5G and 6G. A schematic illustration in **Figure 1** depicts the mutual coupling phenomenon influenced by neighboring antennas' radiation. Impedance mismatch resulting from mutual coupling can lead to scan blindness or total reflection [1–3]. In certain extreme scenarios, the impedance mismatch can lead to total reflection, characterized by a reflection coefficient ($|\Gamma|$) equal to 1.

Mutual coupling significantly affects the radiation properties of individual elements in an array, altering their radiation patterns and characteristics due to electromagnetic field interactions with neighboring elements. This phenomenon induces various effects:

Beam Steering and Shaping: Neighboring elements influence the direction and configuration of energy beams, altering the beam's directionality by modifying the effective electrical length of radiating elements.

Gain and Efficiency Adjustment: Mutual coupling influences the gain and efficiency of individual elements, leading to variations in overall array gain due to changes in radiation efficiency.

Side Lobes and Nulls: Mutual coupling introduces undesired side lobes and nulls in radiation patterns due to constructive and destructive interference between proximate elements.

Impedance Matching: Mutual coupling can cause significant impedance mismatch, affecting the input impedance of radiating elements and impacting power transfer

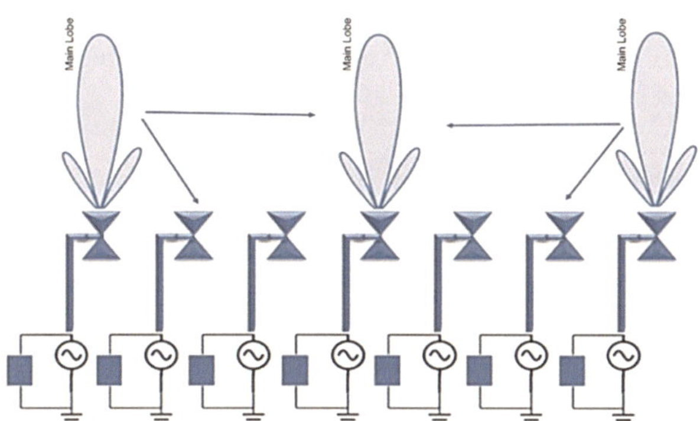

Figure 1.
Mutual coupling phenomenon.

efficiency and radiation properties. Understanding and mitigating mutual coupling effects are crucial for achieving desired array characteristics. Various techniques are available for approximating mutual coupling effects in planar antenna arrays:

During initial design phases, mutual coupling influence may be disregarded, but this can lead to unexpected performance degradation and is unsuitable for manufacturing-ready designs.

Comprehensive simulation of the entire array structure considers practical constraints like mutual coupling using methods such as finite element analysis (FEM) or finite difference time domain (FDTD). However, it becomes computationally intensive for larger arrays.

Infinite Array Simulation approach simulates a single antenna element within an infinite array using periodic boundary conditions. It incorporates mutual coupling effects through Floquet modes but may have limitations in modeling nonuniform or sparse arrays and (ultra)wideband scenarios. It is essential to recognize these methodologies' strengths and limitations to effectively model mutual coupling in antenna array design.

3. Theory of antenna mutual coupling reduction

Arrays have historically been viewed as non-interacting, perfectly matched entities with the feed network. However, in practical settings, interactions among array elements result in mutual coupling, impacting currents, impedances, and element distribution, thus altering the array pattern. This section explores the effects of mutual coupling on impedance and pattern, discussing methodologies for assessing array element impedance and pattern while considering mutual coupling. Hannan's foundational work [4] provides terminology guidance, although alternative terms exist. Passive arrays use a single generator followed by power dividers and phase adjustment devices, while active arrays employ a separate generator for each element, a more common model. The impedance of an isolated element is termed isolated or antenna impedance, while in the array environment with all elements active, it's called active or driving impedance. Measuring active impedance requires full array activation, often achieved through simulation. Element impedance is typically characterized by exciting one element while terminating others passively, usually with a reference load like 50 Ω. Passive impedance closely approximates active impedance due to similar coupling and termination conditions, so "active impedance" often substitutes "passive impedance." For fully active arrays, the term "active impedance in a fully active array" distinguishes cases involving full array activation.

3.1 Impedance impact of mutual coupling

The influence of mutual coupling on impedance manifests through three principal mechanisms, illustrated in **Figure 2a**. The first mechanism involves direct space coupling between the elements within the array, where the proximity of elements results in mutual influence. The second mechanism, indirect coupling, arises from scattering effects induced by nearby objects, such as a support tower. Lastly, the feed network connecting the elements in the array acts as a conduit for coupling. In practical array configurations, measures are taken to mitigate feed network coupling by ensuring proper impedance matching at each individual element. This approach allows the treatment of each element as independent generators, as depicted in **Figure 2b**.

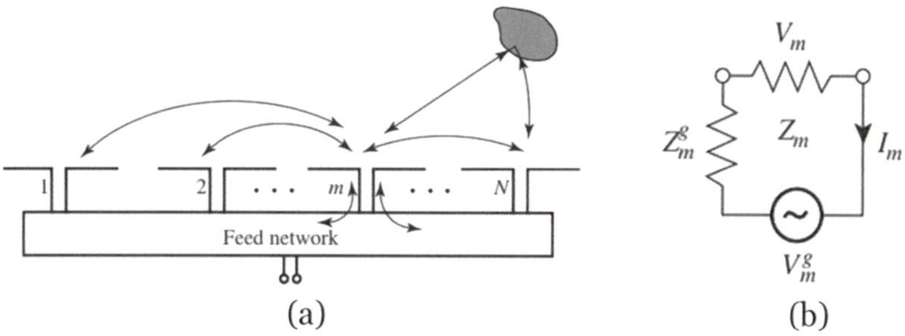

Figure 2.
Mutual coupling in a fully excited array antenna. (a) Mechanisms for coupling between elements of an array. (b) Model for mth element in an array.

In this representation, the mth element is characterized by an applied generator voltage (V_m^g) and terminal impedance (Z_m^g). It is crucial to note that the voltage (V_m) and current (I_m) at the element terminals encompass all the effects of coupling. For the analysis of an array comprising N elements, conventional circuit analysis is employed, treating the array as an N-port network. This methodology facilitates a comprehensive exploration of the impedance effects stemming from mutual coupling, providing valuable insights for both array design and performance evaluation:

$$\begin{aligned} V_1 &= Z_{11}I_1 + Z_{12}I_2 + \ldots + Z_{1N}I_N \\ V_2 &= Z_{12}I_1 + Z_{22}I_2 + \ldots + Z_{2N}I_N \\ &\vdots \\ V_N &= Z_{1N}I_1 + Z_{2N}I_2 + \ldots + Z_{NN}I_N \end{aligned} \tag{1}$$

In the examination of array antennas, the delineation of mutual impedance proves to be a pivotal aspect. The impressed current (I_n) and voltage (V_n) in the nth element, alongside the self-impedance (Z_{nn}), when all other elements remain open-circuited, assume a substantial role in this analysis. Additionally, the mutual impedance between the two terminal pairs of elements, denoted as Z_{mn} (equivalently Z_{nm} due to reciprocity), encapsulates crucial information. To establish the definition of mutual impedance, one must consider a scenario where the initial terminal pair is open-circuited, with the remaining terminals similarly open-circuited. In this prescribed configuration, the mutual impedance Z_{nm} is derived by dividing the open-circuit voltage generated at the first terminal pair by the current supplied to the second terminal pair. This formulation facilitates an accurate characterization of the interaction among individual elements within the array, systematically considering the effects induced by mutual coupling. A comprehensive understanding of mutual impedance equips designers with valuable insights into the array's behavior and performance, thereby facilitating the optimization of its design and radiation properties:

$$Z_{mn} = \left.\frac{V_m}{I_n}\right|_{I_i=0 \text{ for all } i \text{ except } i=n} \tag{2}$$

The precise calculation or measurement of mutual impedance between antenna elements poses a considerable challenge. However, the advent of commercial simulation codes grounded in moment methods has significantly streamlined the evaluation of mutual impedance values, particularly in the context of wire antennas.

These simulation tools offer a convenient means of acquiring reliable results. In measuring mutual impedance between two antennas, a general approach is adopted, extendable to determine mutual impedance between any two elements within an arbitrary array, denoted as Z_{mn}. The specific methodology employed for measuring mutual impedance may vary based on the type of antennas and the experimental setup in use. Leveraging simulation tools and refined measurement techniques enhances the feasibility of determining mutual impedance. This capability provides antenna designers and engineers with invaluable insights into the intricate interactions and coupling effects prevalent among antenna elements.

This knowledge holds paramount importance for the optimization and enhancement of performance in antenna arrays. When an antenna operates in isolation within free space, characterized by a voltage V_1 and a current I_1, the input impedance is a critical parameter to consider for effective system design and integration:

$$Z_{11} = \frac{V_1}{I_1} \textit{ single isolated element} \tag{3}$$

When a secondary antenna is introduced in close proximity to the first, the radiated electromagnetic field from the initial antenna induces currents within the second antenna. Consequently, the second antenna reciprocates by emitting radiation, and this emitted radiation influences the current distribution on the original antenna. The second antenna may be either actively excited or remain unexcited, functioning in a parasitic manner. In either scenario, the second antenna sustains a terminal current denoted as I_2. Subsequently, the aggregate voltage at the terminals of the first antenna is influenced by this induced current. This interaction highlights the intricate coupling effects that arise when multiple antennas operate in close quarters. Understanding these effects is crucial for designing antenna systems that operate efficiently and predictably, especially when considering scenarios involving mutual coupling and electromagnetic interference

$$V_1 = Z_{11}I_1 + Z_{12}I_2 \tag{4}$$

Similarly, the voltage at the terminals of the second antenna is expressed by

$$V_2 = Z_{21}I_1 + Z_{22}I_2 \tag{5}$$

Note that (1) is a generalization of (4) and (5). Now, suppose the second antenna has a load impedance Z_2^g across its terminals $\left(V_2^g = 0\right)$ such that $V_2 = -Z_2^g I_2$. We can write (5) as

$$-Z_2^g = Z_{21}I_1 + Z_{22}I_2 \tag{6}$$

Solving for I_2 and using $Z_{21} = Z_{12}$, we obtain

$$I_2 = \frac{-Z_{21}I_1}{Z_{22} + Z_2^g} = \frac{-Z_{12}I_1}{Z_{22} + Z_2^g} \tag{7}$$

Substituting this into (4) and dividing by I_1, we find that

$$\frac{V_1}{V_1} = Z_1 = Z_{11} - \frac{(Z_{12})^2}{Z_{22} + Z_2^g} \tag{8}$$

The input impedance of the system can be effectively described by incorporating self-impedances (Z_{11} and Z_{22}), (mutual impedance Z_{12}), and the load impedance Z_2^g at the unexcited terminals of antenna 2. In light of the preceding discussion, a schematic representation akin to **Figure 3** emerges as an apt equivalent circuit for delineating the coupling dynamics between two resonant antennas. In scenarios where a single antenna operates in isolation, with antenna 2 positioned at a substantial distance, the mutual impedance Z_{12} diminishes to zero $Z_{12} = 0$Consequently, Eq. (8) simplifies the input impedance to the self-impedance of antenna 1, resulting in $Z_1 = Z_{11}$. Conversely, when antenna 2 is open-circuited in $Z_2^g = \infty$, Eq. (8) indicates $Z_1 = Z_{oc} = Z_{11}$. The act of open-circuiting antenna 2 leads to a cessation of current along its length, a scenario particularly applicable to certain antenna types like half-wave dipoles, where open-circuiting annuls their resonant behavior. However, some antenna types, such as full-wave dipoles, may still exhibit induced current despite being open-circuited, warranting the removal of the second antenna.

To analyze the impedance characteristics comprehensively:

1. Open-circuit (or remove) antenna 2 and measure the open-circuit impedance $Z_{oc} = Z_{11}$ at the terminals connected to antenna 1. In cases involving identical antennas, Z_{22} is equivalent to Z_{11}

2. Short-circuit antenna 2 and measure the short-circuit impedance Z_{sc} at the terminals connected to antenna 1.

3. Utilize the obtained values to calculate the mutual impedance Z_{12}.

4. This systematic approach facilitates a thorough understanding of the coupling effects between resonant antennas, enabling precise impedance characterization for effective antenna design and performance optimization:

$$Z_{12} = \sqrt{Z_{OC}(Z_{OC} - Z_{SC})} \tag{9}$$

The derivation of mutual impedance, as an extension of Eq. (3) with $Z_2^g = 0$, involves the introduction of a generator at the input of one antenna. This process entails determining the voltage at the input terminals of the second antenna, followed by the calculation of the voltage-to-current ratio. Scattering parameters emerge as a widely embraced and favored tool due to their direct correlation with measurements, allowing for facile acquisition through network analyzers. The diagonal elements within the scattering matrix align with the reflection coefficients at each port, thereby furnishing valuable insights into the reflected waves at individual ports:

$$S_{nm} = \Gamma_m \tag{10}$$

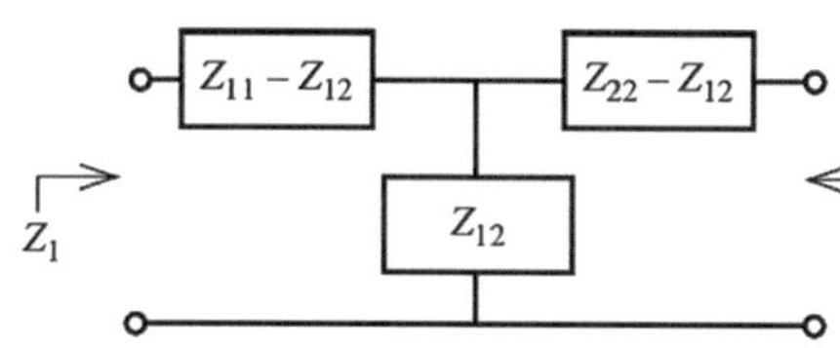

Figure 3.
Network representation of the coupling between two antennas.

In tandem with the impedance matrix, the scattering matrix manifests symmetry in the context of a reciprocal device, signifying that the entry S_{nm} is equivalent to S_{nm}. The non-diagonal components within the scattering matrix furnish essential details pertaining to the coupling between diverse ports. To quantitatively express the coupling between ports m and n in decibels (dB), we introduce C_{mn} as follows:

$$C_{mn} = 20\log|S_{mn}|\ [dB] \tag{11}$$

The entries of the scattering matrix are derived from the impedance matrix, denoted as Z_{mn}. Consider a two-element array where the elements are interconnected by transmission lines with a characteristic impedance of Z_O. For the first element of the array, the scattering matrix entries are given by:

$$S_{11} = \frac{(Z_{11} - Z_0)(Z_{22} + Z_0) - Z_{12}^2}{(Z_{11} - Z_0)(Z_{22} + Z_0) - Z_{12}^2} \tag{12}$$

$$S_{12} = \frac{2Z_0 Z_{12}}{(Z_{11} - Z_0)(Z_{22} + Z_0) - Z_{12}^2} \tag{13}$$

The validity of these equations can be confirmed through straightforward checks. In the absence of mutual coupling ($Z_{12} = 0$), Eq. (12) simplifies to $S_{11} = \Gamma_1 = (Z_{11} - Z_O)/(Z_{11} + Z_O)$, which aligns with the conventional formula for the one-port reflection coefficient. Similarly, when there is no mutual coupling, Eq. (13) yields $S_{12} = 0$, as expected.

Now, it becomes feasible to compute the input impedance of an element in the array using mutual impedance values. For the mth element, the input impedance Z_m can be determined using Eq. (1):

$$Z_m = \frac{V_m}{I_m} = Z_{m1}\frac{I_1}{I_m} + Z_{m2}\frac{I_2}{I_m} + \ \ldots\ + Z_{mN}\frac{I_N}{I_m} \tag{14}$$

The representation of the input impedance, encompassing all constituent elements, denotes the active impedance within a fully excited array. This inclusive characterization takes into consideration the collective influence of coupling effects, wherein mutual impedances are acquired through numerical simulations or measurements employing Eq. (9). The formula explicitly elucidates that the input impedance is shaped not only by the mutual impedances but also by the terminal currents of each individual element. It is essential to underscore that the phase of the terminal currents, meticulously adjusted for pattern scanning, exerts a substantial influence on the input impedance. Consequently, the input impedance of all elements within an array undergoes alterations as the scan angle is adjusted. This dynamic interplay between mutual impedances, terminal currents, and phase adjustments underscores the complexity of the input impedance landscape in array antennas. Understanding these nuances is pivotal for antenna designers and engineers aiming to optimize the performance of array configurations. The intricate relationship between input impedance and scan angle adjustments necessitates a nuanced approach to antenna array design, ensuring adaptability and reliability across a range of operational scenarios.

3.2 Evaluating array pattern with mutual coupling considerations

The examination of mutual coupling's influence on the radiation characteristics of an array constitutes a crucial facet discussed in this section. While a thorough analysis involving simulations or measurements of individual embedded element antennas can provide a comprehensive array characterization, such an approach is often time-intensive and infrequently employed. In lieu of this, approximate techniques are commonly adopted to encompass the effects of mutual coupling, a methodology we shall delineate. Various approaches exist for modeling arrays, ranging from explicit consideration of all mutual impedances to integrating coupling effects into the isolated-element pattern through current adjustments or into the element pattern using the active-element pattern approach. In this discourse, we espouse the active-element pattern approach.

Within the active-element pattern approach, the total array pattern integrates all coupling effects by adapting the excitations of individual elements. This methodology streamlines the analysis, offering a pragmatic balance between accuracy and computational efficiency in studying the impact of mutual coupling on the radiation characteristics of antenna arrays:

$$F_{un}(\theta.\phi) = g_i(\theta.\phi)\sum_{m=1}^{N} I_m e^{j\xi m} \tag{15}$$

In the given context, ξ_m represents the cumulative phase contribution, typically referenced to the center of the array, resulting from spatial phase delay. In the absence of coupling effects, the currents I_m vary proportionally with the excitation voltages:

$$I_m = \frac{V_m^g}{Z_m^g + Z_m} \tag{16}$$

This phenomenon is commonly referred to as "unconstrained excitation" because the voltage at the element terminals varies with the scan angle. Instead of using a constant-voltage feed, a constant-incident power feed is employed. For an infinite array composed of identical elements arranged in a uniform grid, where all elements experience the same mutual coupling, each element is exposed to an identical environment. This implies that the active impedances are uniform across all elements, resulting in all $\{Z_m\}$ being equal to Z_A:

$$I_m = \frac{V_m^g}{Z^g + Z_A} \infty V_m \tag{17}$$

This scenario is applicable to large arrays with equal spacing between elements. However, for finite arrays, the conventional implementation of Eq. (15) involves approximating Eq. (17). This approximation neglects variations in terminal currents caused by mutual coupling and only considers variations in the generator voltage. Obtaining accurate current information for evaluating Eq. (15) is challenging, leading to the common usage of the active-element pattern method, which is discussed in the subsequent sections. In the active-element pattern approach, all coupling effects are taken into account through the active element. The active-element pattern $g_{ae}^n(\theta.\phi)$ is obtained by exciting only the nth element while loading all other elements with the generator impedance Z^g. The active-element pattern results from the direct radiation of the nth element combined with the fields reradiated from other elements, which receive their power through spatial coupling from element n. The formulation of the array pattern in this context is as follows:

$$F_{un}(\theta.\phi) = \sum_{n=1}^{N} g_{ae}^{n}(\theta.\phi) I_n e^{j\xi n} \tag{18}$$

The provided text discusses the correlation between currents and excitation voltages in an array, integrating mutual coupling effects into the active-element patterns. To accommodate potential gain variations, the levels of active-element patterns are referenced to a central element within the array. Although measuring active-element patterns for each array element could be time-intensive, it is often unnecessary. In the context of a large array with uniform elements and equal spacing, each element encounters a similar environment with its nearest neighbors, excluding edge elements. Hence, it is customary to approximate the factorization of the active-element pattern using an average pattern denoted as $g_{ae}^{n}(\theta.\phi)$, representing the normalized pattern for a typical central element in the array:

$$F_{un}(\theta.\phi) = g_{ae}(\theta.\phi) \sum_{n=1}^{N} I_n e^{j\xi n} \tag{19}$$

The notable advantage of this approach lies in its utilization of the summation within the equation, based on the array factor derived from basic theory, independent of mutual coupling considerations. Furthermore, all coupling phenomena are encompassed within the average active-element pattern, determinable through a single measurement of the pattern exhibited by a central element within an extensive array.

Within the domain of mathematical approximations, the normalized rendition of Eq. (19) yields a crucial and noteworthy outcome:

$$F(\theta.\phi) = g_{ae}(\theta.\phi) f(\theta.\phi) \tag{20}$$

Undoubtedly, this formulation adopts a semblance to the pattern multiplication formula, denoted as Formula (20), yet intricately interwoven within it are the ramifications of mutual coupling. This approximation is frequently employed in practical applications. The average pattern of the active element assumes a pivotal role in the construction of arrays. Prior to the creation and testing of the grand array, replete with active elements, the pattern of an individual active element surrounded by passive elements undergoes measurement and evaluation. If the pattern of the active element proves to be irregular, so too will be the grand array. It has been demonstrated that the active input impedance can be deduced from the data of the active-element pattern [5]. The pattern of the isolated element differs from those of the active elements within the array. However, in an array of sufficient magnitude, the average pattern of the elements represents the majority of the active-element patterns within the array. The question of how extensive an array must be before (20) can be applied holds great significance and is addressed through guidelines and examples. As the size of the array increases in magnitude (ideally, an infinite multitude of elements), the active-element patterns of individual elements tend toward becoming identical and aligned with the average pattern of the active element. Elements near the borders of a finite array bear patterns that deviate from the pattern of the central elements, which should closely adhere to the average-element pattern. This is evident in the patterns depicted in **Figure 4**, illustrating the active-element patterns of each element in an array of eight microstrip patch antennas spaced apart by 0.57λ. The

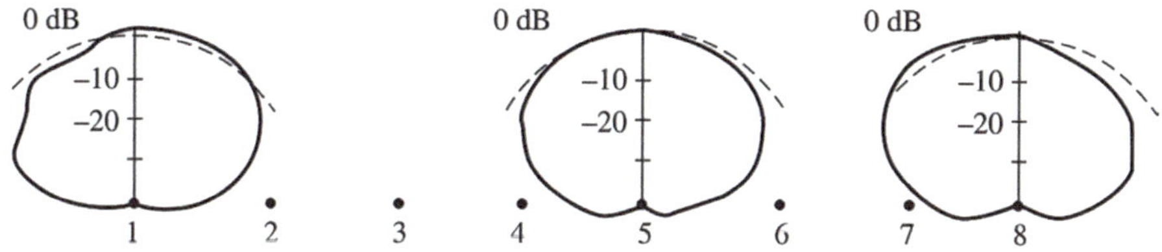

Figure 4.
Measured active-element patterns for three elements, an interior element (5) and the end elements ((1) and (8)), of a linear array of eight microstrip patch elements spaced 0:57lapart.

polar -dB patterns displayed for selected elements were measured while that element was excited and all others inactive. The patterns of elements 2 through 7 (only pattern 5 is shown) are nearly indistinguishable and symmetrical. The patterns of the edge elements ((1) and (8)) are distorted owing to the asymmetrical environment of the array and the finite effects of the ground plane's borders. Note that the greatest distortion in the patterns occurs on the side away from the array. Similar findings have been reported concerning dipoles above a ground plane.

The overarching pattern of the array undergoes perturbation due to the inherent coupling within. Arrays are crafted with excitation voltages tailored to our specifications and modeled with voltage generators. In the absence of coupling, the currents at the terminals maintain proportionality to the excitation voltage, and the desired pattern emerges, as implied by Formula (17). However, when coupling emerges, the pattern undergoes deformation. It often occurs that the terminal currents deviate significantly from the desired excitation, yet the pattern remains largely unaffected. Coupling not only influences the pattern but also impacts the gain and polarization of the entire array pattern. Mutual coupling does not alter the distribution of currents on elements within the array, but it does change the complex currents at the terminals. Compensation schemes can anticipate how coupling will alter the terminal currents and adjust the excitation voltages to elicit the desired currents. Moreover, physical methods exist to directly mitigate mutual coupling, such as the insertion of conductive baffles and resembling fences, among the elements.

4. Methods of reducing mutual coupling

Mutual coupling reduction refers to techniques and methods employed to minimize the effects of mutual coupling in antennas and other electronic systems. Here are a few common methods used to reduce mutual coupling: Increased spacing: One of the simplest ways to reduce mutual coupling is by increasing the physical distance between antennas or components. By spacing them further apart, the coupling effect diminishes, resulting in reduced interference. Alternating structures: Alternative structures are two or three-dimensional arrays of elements that are alternately placed together. Types of these structures include frequency-selective surfaces, crystal photonics, and metamaterials. The analysis of these structures is done using simple Floquet analysis. EBG (electromagnetic bandgap): EBG structures are highly desirable in modern antenna and microwave designs. In essence, EBG structures are artificial periodic arrangements of dielectric materials and metallic conductors. They exhibit the unique property of preventing radiation and facilitating the propagation of electromagnetic waves within a specific frequency range and for all angles. EBG structures can effectively suppress unwanted radiation and support desired polarization modes. Mutual coupling reduction involves techniques to minimize its effects in antennas and electronic systems. Common methods include:

- *Increased spacing:* Widening the physical distance between antennas reduces coupling and interference.
- *Alternating structures:* Using structures like frequency selective surfaces (FSS), crystal photonics, and metamaterials can reduce coupling effects.
- *EBG (electromagnetic bandgap) structures:* Artificial periodic arrangements of dielectric materials and conductors, EBGs prevent radiation and facilitate electromagnetic wave propagation within specific frequency ranges.
- *FSS (frequency selective surfaces):* Made of alternating patches on a grounded dielectric layer, FSS act as specialized filters, absorbing specific frequencies.
- *PBG (photonic bandgap) structures:* These reduce surface waves, improving antenna efficiency, bandwidth, and reducing interference.
- *Fractals:* By exploiting space-filling properties, fractals reduce size and cross-coupling in microarrays.
- *DGS (Defected Ground Structure):* Meandering techniques in the ground plane reduce mutual coupling by creating gaps perpendicular to the feeding point.
- *Resonator rings:* Used for protection and enhancing radiation, resonant levels create frequency gaps, reducing cross-coupling.
- *Geometric changes to the patch:* Altering patch geometry, like using fractal patches, reduces mutual coupling. The choice of technique depends on factors like frequency range, available space, cost, and performance objectives. Future methods for mutual coupling reduction will be explored.

4.1 Reduction of mutual coupling with liquid dielectric

Numerous techniques implemented thus far aimed at mitigating mutual coupling have resulted in the isolation of sizable elements, unwieldy structures, intricate numerical computations, and irreversible structural modifications. The focal point of this research lies in the design of array antennas tailored for contemporary, practical, and multidisciplinary applications. Consequently, this study proposes a pragmatic approach for segregating antennas through the utilization of water. This innovative and controllable method circumvents the need for complex numerical equations, validating the structural efficacy via the wave scattering method. The procedure involves minimizing mutual coupling between two elements by introducing a liquid dielectric, specifically distilled water, between the components of an antenna array. The future potential of this technique to suppress crosstalk holds promise for diverse applications such as the integration of solar panels and energy harvesting with electromagnetic structures [6, 7], transparent electromagnetic structures [8], and low radar cross-section (RCS) devices [9–11]. The ensuing sections outline the primary specifications of this device, emphasizing its applicability in air and marine environments:

- Utilize liquid water to mitigate cross-coupling between diffusing components.
- The method is fully implementable due to the confined glass plexiglass structure.
- Strongly recommend optically transparent solar panel integration solutions.
- The proposed structure enhances the characteristics of the primary antenna.

Following this, a comprehensive discussion ensues, covering the presented structure, including a summary of water characteristics concerning electromagnetic (EM) waves, the design of a simple microstrip antenna array, and a cross-coupling reducer featuring a confined water device. The subsequent section presents the results obtained from numerical, experimental, and analytical solutions.

4.1.1 Wave propagation in water

Owing to the elevated conductivity loss of seawater in the mid-RF range's propagation of EM waves, the normal conductivity is established to be 4 S/m [12]. In contrast, freshwater exhibits a conductivity of 0.01 S/m, which is 400 times lower than that of seawater. Consequently, EM wave propagation in freshwater may prove more effective than in seawater. This section delves into the theoretical aspects of water absorption ability. As EM waves transition from open air to freshwater, two types of losses warrant investigation:

- Connection loss, arising from air reflection and water interference.
- Propagation loss of EM waves in water, explored based on water material properties.

This loss is harnessed as a novel technique to diminish undesirable coupling between two propagation components. Both loss mechanisms are analytically defined in this section, drawing from the scattering presentation depicted in **Figure 5**. Eq. (1) is used to measure total propagation loss of the structure where (α_p) and (α_t) give the propagation and transmission losses:

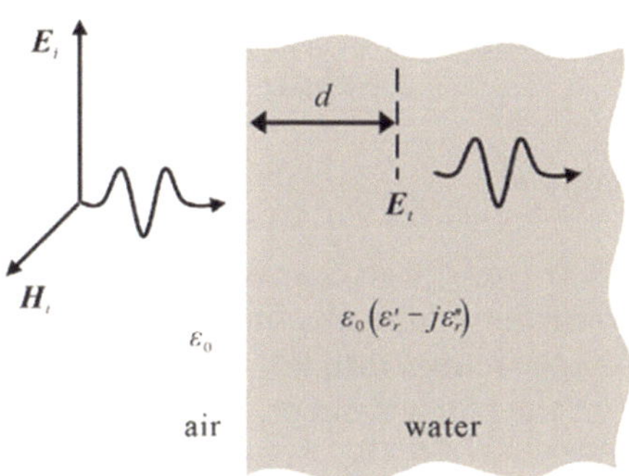

Figure 5.
Penetration plan of EM waves into freshwater.

$$\alpha_{total} = \alpha_t + \alpha_p = 10 \log\left(\frac{P_t}{P_i}\right) \tag{21}$$

$$T = \frac{2\eta_1}{(\eta_0 + \eta_1)} \tag{22}$$

where a, d, T, η0, and η1 denote the imaginary component of the wave propagation constant, depth of propagation in water, transmission coefficient, and air and water characteristic impedances, respectively. In addition, EM wave interaction with water can be modeled with the Debye dispersive dielectric constant, as in Eq. (7):

$$\overline{\varepsilon_r}(f) = \varepsilon_\infty + \left(\frac{\varepsilon_s - \varepsilon_\infty}{1 + \left(i\frac{f}{f_{ref}}\right)}\right) - \frac{i\sigma}{2\pi f \varepsilon_0} \tag{23}$$

It can be understood that the propagation losses are determined by the propagation depth and water permittivity. Therefore, the surface current transmission from one patch to another may be prevented by applying a proper material and geometry. According to the total loss equation, the water depth is chosen so that it only blocks the transmission of surface currents from the first to the second patch while causing no interference with the propagation of antenna. As a theoretical proof of concept, the wave dissipation value from Eq. (1) can be used to propose water as a novel mechanism for mutual coupling reduction.

4.1.2 Antenna structure description

This section introduces a novel approach aimed at mitigating mutual coupling in a microstrip antenna array through the utilization of a liquid dielectric, specifically distilled water. The proposed antenna array is configured as a two-element microstrip patch array, engineered for operation at 7 GHz, corresponding to the C-frequency range. The foundational antenna structure employs a typical geometry, serving as a baseline to illustrate the coupling characteristics inherent to an array antenna. To further explore the potential of reducing antenna mutual coupling, a structured water medium is introduced into the array. This serves as a demonstrative application of a transparent dielectric, strategically employed to diminish mutual coupling within the antenna system. This innovative integration of a structurized water element represents a promising avenue for enhancing the performance and applicability of microstrip antenna arrays, particularly in scenarios where mutual coupling reduction is critical.

4.1.3 Base antenna configuration

The base array antenna is fed by two coaxial Sub-Miniature version A (SMA) connectors. These coaxial cables exhibit an outer conductor linked to the ground plane, with a center conductor routed through a hole in the ground plane, establishing electrical connectivity with the patch element. The antennas within the array are strategically spaced at a distance of $0.4\lambda_0$. Among various array configurations, the adjacent 1×2 array stands out with the highest coupling value, rendering it highly compatible with circuit board technology. **Figure 6** provides a visual representation of the planned array, while **Table 1** outlines the specific specifications of the antennas.

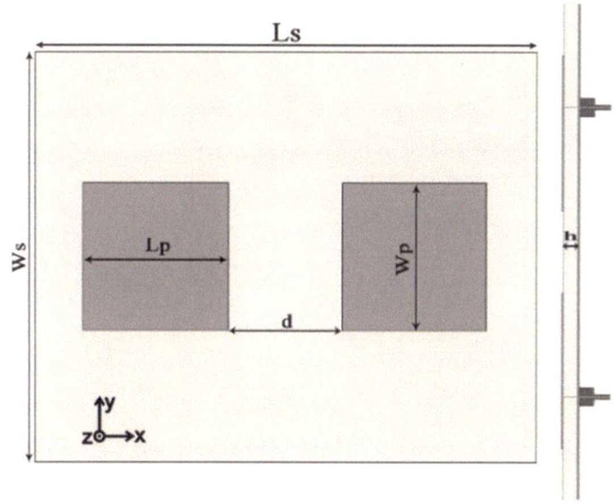

Figure 6.
Top and side views of the antenna before inserting the coupling reduction structure.

Ls	Ws	h	d	Lp	Wp	Ls	Ws
100 mm	80 mm	1.6 mm	23.25 mm	28.75 mm	28.75 mm	100 mm	80 mm

Table 1.
Dimensions of the antenna are shown in ***Figure 6****.*

The substrate's thickness, relative permittivity, and loss tangent are designated as 1.6 mm, 4.8, and 0.025, respectively. A comprehensive analysis of the scattering parameters diagram is conducted both before and after the incorporation of the coupling reduction structure involving a liquid dielectric. This analysis aims to quantify the coupling effect between the two antenna components. Moreover, the simulation yields a maximum radiation gain of 10.7 dB for the antenna without a liquid dielectric at the resonance frequency of 7 GHz. This information contributes to a thorough understanding of the antenna's performance characteristics in the absence of the proposed coupling reduction mechanism.

4.1.4 Mutual coupling reduction mechanism

In order to mitigate mutual coupling, a novel approach involving a confined water structure on the base antenna is presented. Plexiglass material is employed to construct the structure, exhibiting specific characteristics suitable for mounting the liquid dielectric between the two antenna components. The choice of plexiglass for this project is motivated by its high resilience to diverse climatic conditions, ready availability in terms of thickness and dimensions, cost-effectiveness, transparency, and minimal impact on wave propagation. For the intended configuration, plexiglass sheets are meticulously assembled into a hollow container capable of accommodating the liquid dielectric, as illustrated in **Figure 7(a)** and **(b)**. Notably, the bottom of the container covering the antenna remains exposed, allowing the liquid dielectric to make direct contact with the surface of the microstrip antenna substrate. When a surface current emerges on the antenna substrate plane, it is absorbed by the liquid dielectric before being radiated. The hollow containers crafted from plexiglass are filled with distilled water, characterized by ε_r, μ_r, and σ values of 90, 1, and 1.59 [S/m],

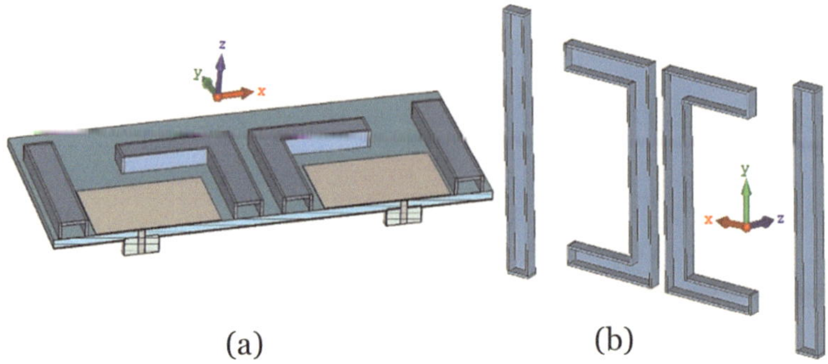

Figure 7.
Illustrates the proposed design for mutual coupling reduction, providing a sectional view of the structure installed over the antenna (a) and depicting the entire plexiglass structure (b).

respectively. The pivotal operation for coupling reduction is executed by the first segment with a ⊏ shape strategically positioned between the two radiating elements of the microstrip antenna array. It is imperative to note that substantial changes in the medium's temperature may impact the effectiveness of this approach under stable coupling reduction conditions. The introduction of water into the array mirrors the principles observed in meander line structures designed to reduce mutual coupling between patch antennas [13]. It is essential to highlight that any alteration in the overall structure of water confinement for coupling reduction would be necessitated if there are new frequency goals to be achieved.

4.1.5 Characterization results

In the transmission of surface current from the first patch to the second, it encounters a semiconductive state within the water medium, incurring some losses. This distinctive feature plays a pivotal role in diminishing the coupling within the antenna array by mitigating the detrimental effects of the surface current transmitted between the two patches. To detail these characteristics, the CST-Microwave Studio software is employed to execute all three stages of design and simulation, as expounded by previous works [14]. The mutual coupling reduction structure, incorporating a liquid dielectric, is strategically inserted between two microstrip patch antennas, as depicted in **Figure 8**.

Stage 1: As previously mentioned, the proposed structure comprises two distinct sections in ⊏ and I forms. An air gap exists between the two ⊏ – form segments. The

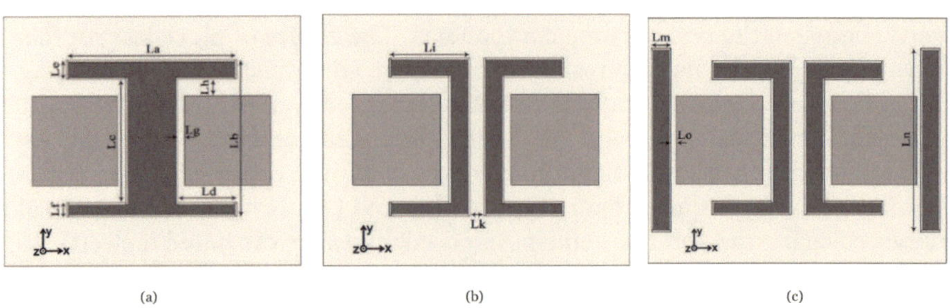

Figure 8.
The structure was designed by applying the air gap. (a) Step 1; (b) step 2; (c) step 3.

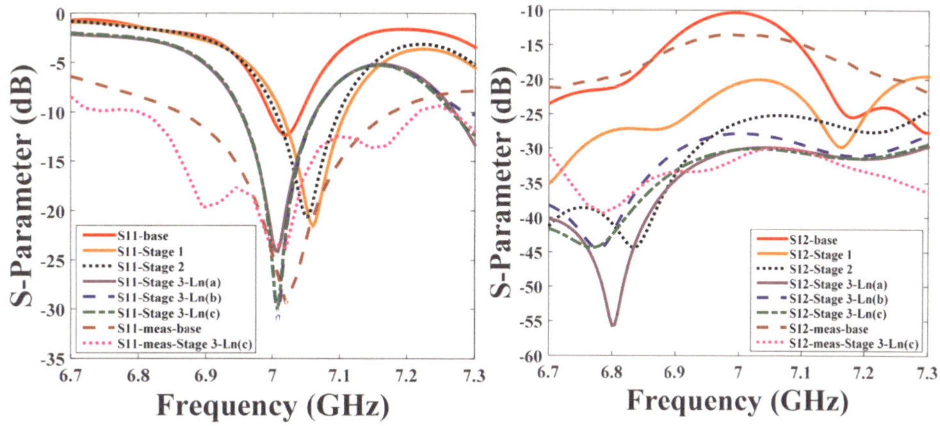

Figure 9.
S11 and S21 scattering parameters of the antenna of sample 3 for different lengths ln.

impact of this air gap is significant, as neglecting it results in the transmission of electromagnetic (EM) waves and surface currents from the left to the right, effectively behaving as a conductor of waves (refer to **Figure 8a**). The scattering parameters are initially evaluated before the introduction of the air gap, as illustrated in **Figure 9**.

Stage 2: The dimensions of the ⊏ – section are substantiated through parametric simulations. In the second stage, the structure, excluding the I-form section, undergoes examination. A 4.4 mm air gap is introduced, as presented in **Figure 8b**.

Stage 3: The final stage incorporates the antenna's I-form structure (**Figure 8c**). Scattering diagrams for various I-form dimensions are depicted in **Figure 9**. A comparative analysis of the scattering parameters for Sample 3 (**Figure 8c**), featuring different lengths (Ln), along with the antenna's radiation pattern, is detailed in **Table 2**. Notably, an increased length of the I-form structure (Ln) correlates with an enhanced radiation pattern. The optimum length of the I-form section that yielded optimal results was determined to be 59 mm. All dimensions of **Figure 8** are listed in **Table 3**, and The comparison of the radiation parameters of the antenna Sample 3 with Ln(c) in the measurement and simulation stages is shown in **Figure 10**.

Stage 3 Ln (mm)	Frequency (GHz)	S11 (dB)	S12 (dB)	Directivity (dB)
Ln(a) = 30	7	−24	−30	8
Ln(b) = 44	7	−30	−28	7.8
Ln(c) = 59	7	−31	−30	10

Table 2.
Variations in S11, S21, and radiation pattern for different ln values.

La	Lb	Lc	Ld	Le	Lf	Lg	Lh
57.2 mm	50.19 mm	39.6 mm	19.8 mm	6.05 mm	4.54 mm	2.83 mm	5.4 mm

Table 3.
*Dimensions of the antenna are shown in **Figure 8**.*

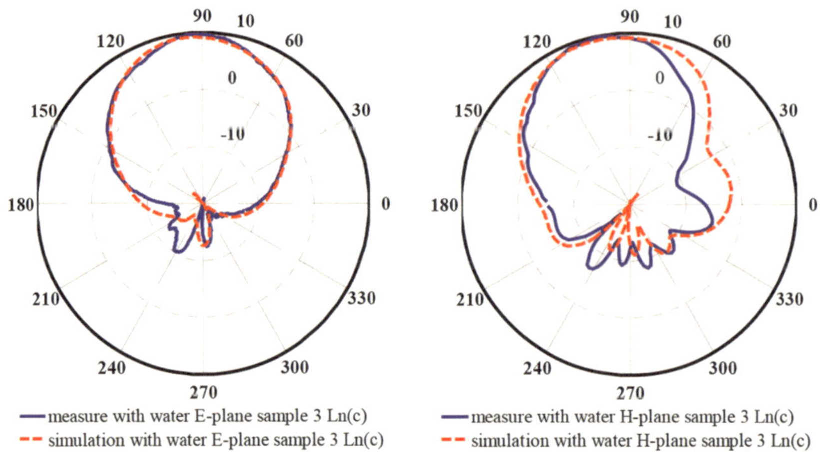

Figure 10.
The diagram radiation parameters as compared in antenna of sample 3 with ln(c) in measure and simulation stages.

All stages of the structure design have been thoroughly examined in the preceding sections. The culmination of these investigations, as evidenced in the last section and elucidated in the analysis presented in **Table 4**, reveals that the most effective configuration for coupling reduction is achieved in Sample 3 with Ln(c). Subsequent analyses delve into the scattering of waves and the dynamics of surface current movement. As previously outlined, this approach mitigates mutual coupling through the absorption and confinement of surface currents. The design methodology involves establishing equilibrium in dielectric coefficients and determining suitable water height and depth. In **Figure 11(a)**, the surface current distribution on the antenna is depicted without the insertion of the coupling reduction structure, while **Figure 11(b)** explicitly illustrates the absorption and confinement of the surface current facilitated by the designed structure. Furthermore, **Figure 12(a)** elucidates the direction of electric field propagation in the antenna lacking the coupling reduction structure. Contrarily, in **Figure 12(b)**, the inhibitory effect of the designed structure on the transmission of electric fields from the first patch to the second patch is evident, effectively creating an obstruction. These visualizations provide valuable insights into the efficacy of the coupling reduction mechanism in redirecting and mitigating surface currents within the microstrip antenna array.

Parameter	Dimensions (mm)	Parameter	Dimensions (mm)
a	14	G	5
b	12	H	1
c	12	K	0.5
d	9	M	2
e	9	N	5
f	1		

Table 4.
*Dimensions of the proposed antenna (**Figure 14**).*

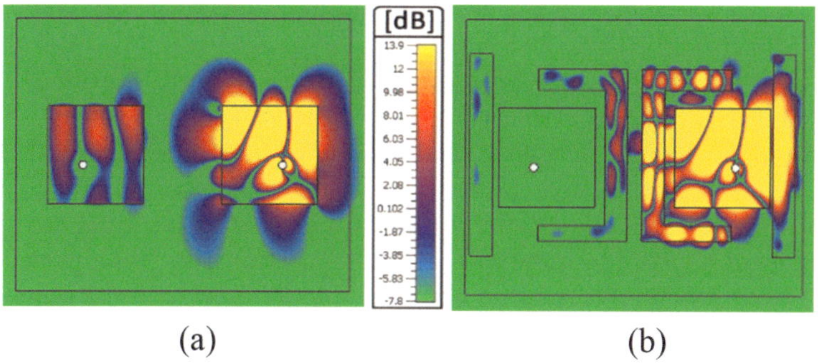

Figure 11.
Distribution of surface current on the designed antenna. (a) the prototype antenna without inserting the coupling reduction structure, (b) the prototype antenna with inserting the coupling reduction structure as sample 3 with ln(c).

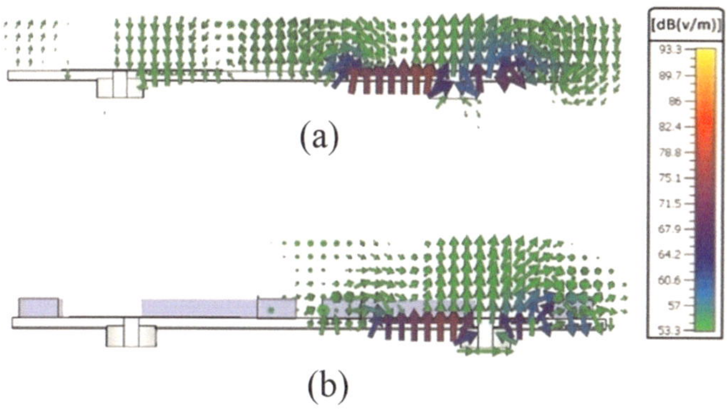

Figure 12.
Distribution direction of electric field propagation in the designed antenna. (a) the prototype antenna without inserting the coupling reduction structure, (b) the prototype antenna with inserting the coupling reduction structure as sample 3 with ln(c).

The constructed prototype of the antenna, featuring the integration of the mutual coupling reduction structure with a liquid dielectric (distilled water), is presented in **Figure 13**. This specific configuration is labeled as Sample 3 with Ln(c), and it is depicted from various viewpoints to provide a comprehensive understanding of its physical attributes.

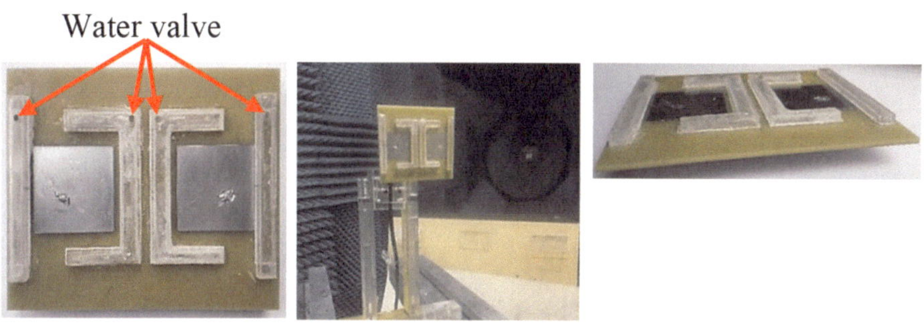

Figure 13.
The antenna sample fabricated with mutual coupling reduction structure with liquid dielectric (distilled water) as sample 3 with ln(c).

4.2 Reduction of mutual coupling with graphite materials, MTM, and EBG design

This method introduces a specialized approach to mitigate cross-coupling by integrating graphite materials, metamaterials (MTM), and electromagnetic bandgap (EBG) design. The proposed antenna is configured as part of an array, with a specific pair intended for testing the structural performance. The array antennas are tailored for use in extensive screens, a crucial consideration for large-scale applications. In dimensions spanning several hundreds, the incorporation of a graphite structure contributes to a reduction in antenna weight due to its inherent lightweight properties. This design not only significantly minimizes mutual coupling losses but also enhances fundamental antenna parameters such as gain and radiation efficiency. These improvements signify an overall enhancement in antenna performance, highlighting the efficacy of the proposed concept for various applications. The investigation begins by examining the relationship between graphite and the propagation of electromagnetic waves. Subsequently, the array antenna is meticulously designed, incorporating the coupling reduction structure. In the final step, the performance of the multiple-input multiple-output (MIMO) antenna is analyzed to validate the effectiveness of the proposed method.

4.2.1 Properties of graphite in the propagation of electromagnetic waves

Graphite sheets are compressible, porous, and electrically conductive materials extensively used in diverse applications, including fuel cells, flow batteries, and electromagnetic shields. The sheet density, a critical parameter influencing electrical conductivity, exhibits a linear relationship with electrical conductivity [15, 16]. Researchers have shown interest in utilizing graphite compounds as effective electromagnetic absorbers due to their electrical conductivity, chemical stability, and ease of manufacturing at a low cost. Recent attention has been directed toward graphene and carbon nanotubes for their remarkable properties, although their impact on electromagnetic wave absorption may be influenced by factors like room temperature carrier mobility [17]. Graphite compounds, characterized by low density, large surface area, high-temperature resistance, oil absorption, and excellent electrical conductivity, present a viable and cost-effective solution for industrial applications. The microwave current behavior in graphite compounds is governed by carrier transfer within the graphite layers, leading to high electrical conductivity and reduced resistance [18]. In line with transfer theory, computational formulas for single-layer adsorbents are expressed in the following equations. These properties make graphite a compelling material for applications requiring effective electromagnetic wave absorption [19, 20]:

$$z_{in} = \sqrt{\frac{\mu_r}{\varepsilon_r}} \tanh\left(j\frac{2\,\pi f d}{c}\right)\sqrt{\mu_r \varepsilon_r} \tag{24}$$

$$\Gamma = \frac{\sqrt{\frac{\mu_r}{\varepsilon_r}} \tanh\left(j\frac{2\,\pi f d}{c}\right)\sqrt{\mu_r \varepsilon_r} - 1}{\sqrt{\frac{\mu_r}{\varepsilon_r}} \tanh\left(j\frac{2\,\pi f d}{c}\right)\sqrt{\mu_r \varepsilon_r} + 1} \tag{25}$$

where ε_r and μ_r represent the relative permittivity and the relative permeability of the adsorbents, respectively. z_{in} is the input impedance normalized because it relates to the impedance in the free space, d is the thickness of the absorber, c is the speed of light, and f is the frequency of the microwave in the free space.

4.2.2 Antenna design

The antenna structure proposed in this study is depicted in **Figure 14**. This microstrip multiple-input multiple-output (MIMO) antenna comprises two identical rectangular radiation patches designed on an FR4 substrate. The substrate has dimensions of 81.7 × 52.85 mm and a thickness of 1.6 mm, possessing a dielectric constant of 4.3. Both antennas are connected to a 50 Ω SMA port. The separation between the two patches is set at 20 mm, a relatively small distance chosen to minimize coupling effects. Increasing this distance excessively could result in a reduction of coupling, but such adjustments are deemed beyond the scope of this study. **Table 5** provides the specific dimensions of the proposed antenna.

4.2.3 Structure description utilizing graphite for mutual coupling reduction in Array antennas

In this research, we propose a structure designed with graphite to alleviate mutual coupling among array antennas, as illustrated in **Figure 14**. Building upon the principles outlined in Section II, where expanded graphite's resistive absorbent nature was detailed, we employ a graphite sheet, a compressible, porous, and electrically conductive material, as the substrate for the proposed structure. To enhance the graphite structure, we incorporate a perfect electric conductor (PEC) electromagnetic bandgap (EBG) layer. The specific graphite utilized in this study has a dielectric constant (εr) of 12, magnetic permeability (μ) of 1, and a thickness of 1 mm. **Figure 15** presents diverse perspectives of the graphite and EBG composite structure.

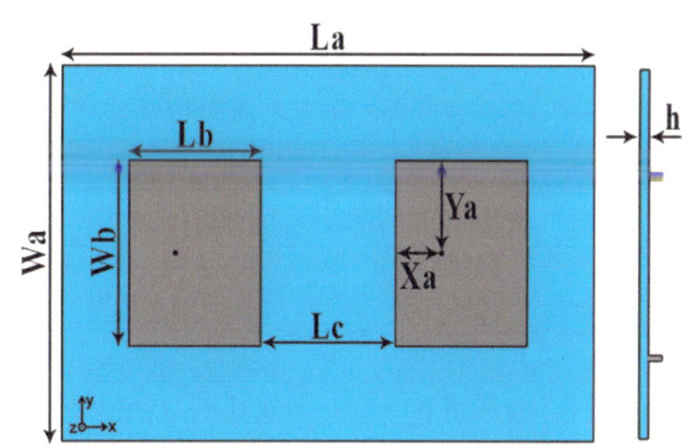

Figure 14.
Sample of the proposed microstrip MIMO antenna.

Parameter	Dimensions (mm)	Parameter	Dimensions (mm)
Wa	53.85	Lc	20.42
Wb	26.93	Xa	7.03
La	81.7	Ya	13.46
Lb	20.43	H	1.6

Table 5.
*Dimensions of the proposed antenna **Figure 14**.*

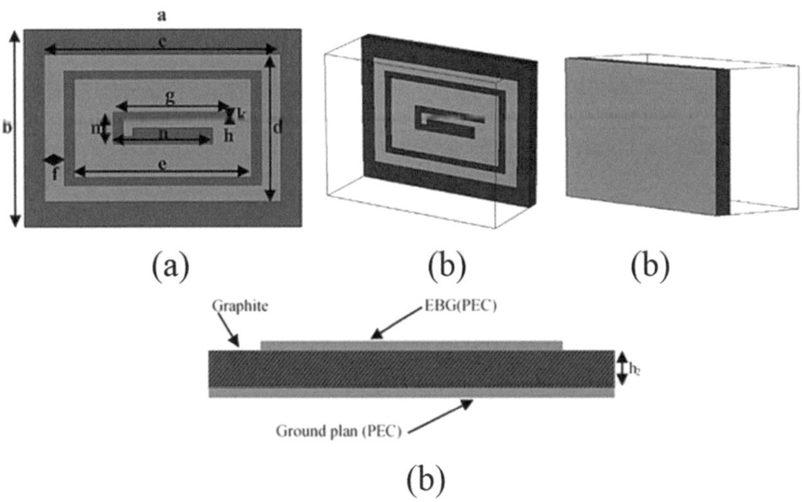

Figure 15.
Schematic of the combined unit structure of graphite and EBG to reduce mutual coupling: (a) general and top view, (b) three-dimensional view back, (c) three-dimensional view of the front of the structure, and (d) side view.

EBG structures offer versatility in configuring stop-bands, pass-bands, and bandgaps. They effectively impede the propagation of electromagnetic surface waves across different frequencies. The EBG structure is composed of periodic elements, either dielectric or metallic, arranged as a single cell and repeated. By adjusting the unit cell's repetition shape and resonance value, bandgaps can be tailored. To analyze the designed structure in **Figure 15** and ascertain the magnetic bandgap, we employ the eigenmode method, considering alternation in both the X and Y axes.

One method for inducing capacitive properties involves placing a rectangular ring around the structure. In our approach, we create capacitive properties in the EBG by encircling the internal structure with a loop and introduce inductive properties by incorporating cuts in the center. By adjusting the structure's dimensions, we determine the optimal values for creating a magnetic bandgap at the desired frequency. **Table 4** provides the dimensions of the proposed graphite structure in **Figure 15**.

Two noteworthy considerations arise: firstly, the utilization of the inner region of the light cone to determine the bandgap, and secondly, the exclusion of the mode 1 curve in the bandgap calculation, as depicted in **Figure 16**. The latter is specific to mode 1, and its exclusion is justified by examining the electric field vectors' symmetry inside the substrate.

As shown in **Figure 16**, the yellow area represents the magnetic bandgap of the structure, ranging from 3.1 GHz to approximately 3.75 GHz. This denotes a magnetic forbidden band within this frequency range, where electromagnetic waves are impeded from passing through the structure. The graphite sheet's structure inhibits surface currents from entering the graphite layers, reducing surface currents at the output. **Figure 17** illustrates the reduction of mutual coupling within the graphite structure and the attenuation of input surface waves, aligning with the schematically designed EBG structure to induce inductance and capacitive properties. The incoming surface currents enter the graphite structure, vibrating between layers and suppressing surface currents, consequently diminishing the outgoing surface current. The designed EBG structure further aids in reducing mutual coupling between antenna radiating patches by creating inductor and capacitor properties and

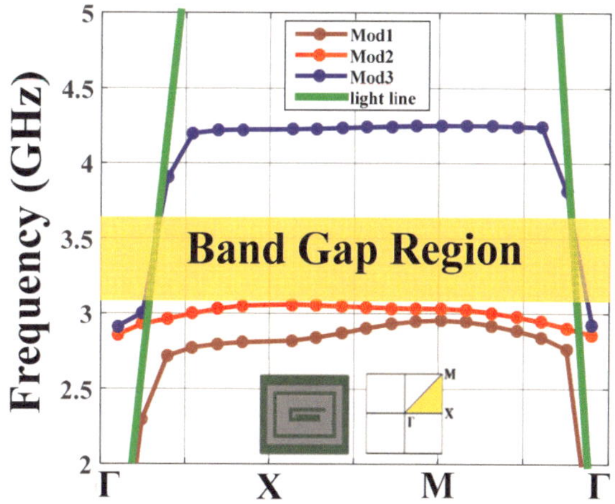

Figure 16.
Showing the range of the magnetic gap band structure from the output of the dispersion diagram.

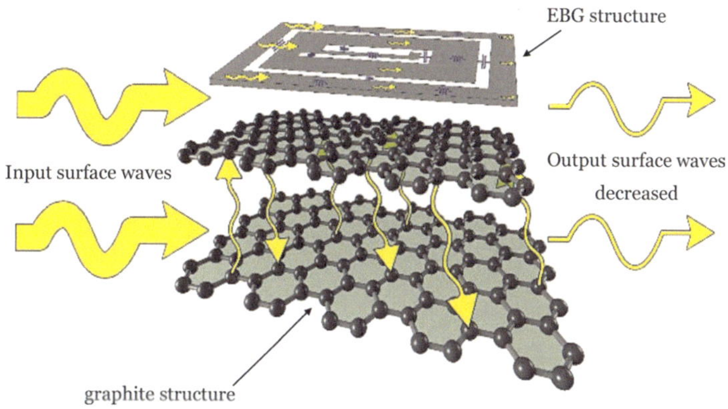

Figure 17.
Showing the effect of graphite structure in reducing mutual coupling.

establishing an electromagnetic forbidden band at the desired frequency. The physical properties of graphite and its interaction with electromagnetic waves were discussed in Section II, emphasizing graphite's conductivity and the impact of dimensions and thickness on the magnetic bandgap frequency, as described by Eq. (1).

4.2.4 Analysis of results

This section delves into the examination and scrutiny of outcomes arising from the integration of the designed electromagnetic bandgap (EBG) structure with graphite amid the radiating elements of the designated antenna. The antenna, featuring coupling reduction structures positioned between the two radiating elements, is visually represented in **Figure 18**. The surface disparity between the antenna ground and the meticulously designed graphite structure is determined to be 0.6 mm, as evident from **Figure 18a** and **b**. This was achieved by strategically cutting the substrate and ground part of the antenna structure, allowing for the insertion of graphite between the two radiator elements. Upon analysis, it is discerned that the proximity of antenna patches

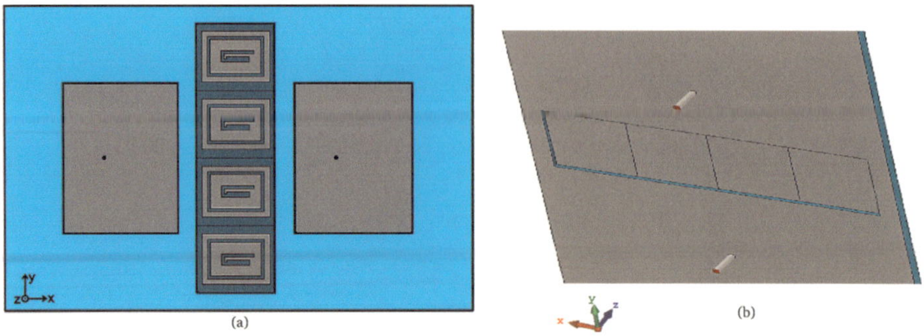

Figure 18.
Antenna with coupling reduction structures between two radiating elements: (a) top view and (b) bottom view.

influences the destructive effects, with a more pronounced impact when the antenna patches are in close proximity. Conversely, as the patches are situated farther apart, the impact diminishes. Consequently, a lower LC (line-of-centers) results in higher mutual coupling, whereas a higher LC leads to diminished mutual coupling. The fundamental objective of the structural design is to curtail mutual coupling within the compact structure, particularly focusing on minimizing the mutual coupling between the two radiation patches positioned at the shortest distance from each other.

In the assessment of the coupling effect between the two elements of the array antenna, a thorough analysis was conducted through the examination of scatter plots of S21 both before and after the incorporation of the graphite structure. **Figure 19** illustrates the comparative diagram, categorizing the graphite structure arrays into four modes: single, double, triple, and quadruple. Notably, the quadruple array emerges as the most suitable configuration, exhibiting the highest degree of mating. For comprehensive insight, the comparative diagram of scattering parameters in **Figure 19**, representing the antenna before and after the integration of the graphite structure, reveals a substantial reduction of 30 dB in coupling. In the context of multiple-input multiple-output (MIMO) systems, achieving a coupling reduction exceeding 15 dB is deemed highly commendable. The synergy between electromagnetic bandgap (EBG) and the graphite substrate is pivotal. The graphite plate contributes significantly to mutual coupling reduction, leveraging its inherent properties that

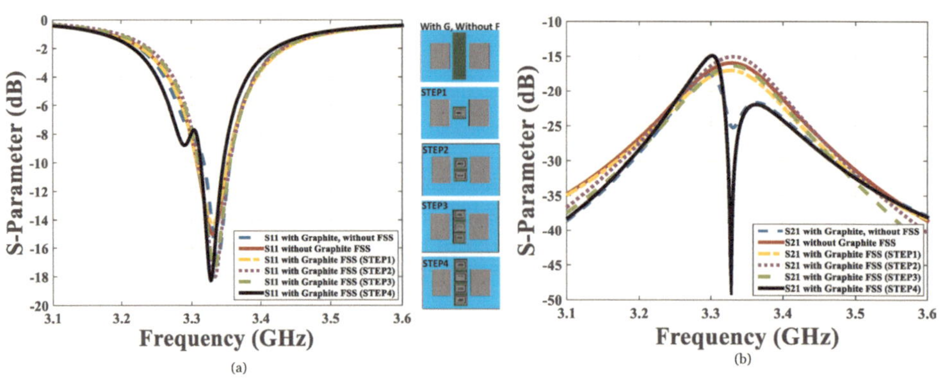

Figure 19.
Antenna scattering parameters: Diagram of without and with graphite structure: (a) S11 and (b) S21 diagram.

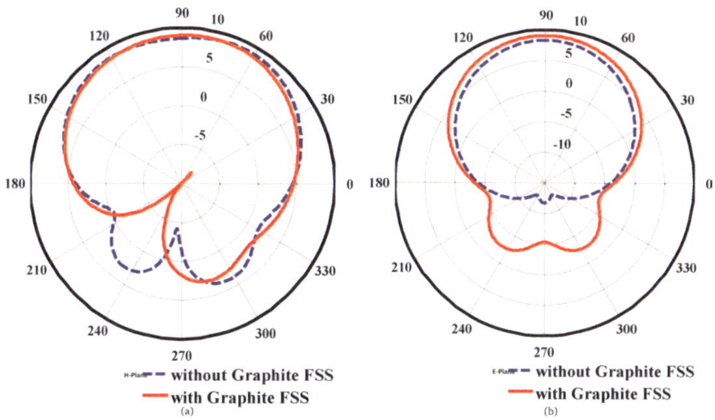

Figure 20.
Diagrams of antenna radiation structure: Comparison without and with graphite structure: (a) H-plane and (b) E-plane.

impede the propagation of electromagnetic waves. Simultaneously, the EBG structure plays a complementary role in reducing mutual coupling. The combined utilization of these two structures in this study yields a more pronounced reduction in mutual coupling. **Figure 19** presents a comparative analysis of the graphical representations, distinguishing between the structure with graphite alone and the structure incorporating both graphite and EBG. The radiation pattern of the antenna, specifically designed for mutual coupling reduction among array elements, is elucidated in **Figure 20**. The comparison chart therein indicates a notable increase in gain, escalating from 8.1 to 9.3 dB, underscoring the efficacy of the integrated structures in enhancing antenna performance.

The assessment of the antenna's radiation efficiency before and after the integration of the graphite structure is presented in **Figure 21**. The comparative analysis in **Figure 21** reveals a substantial improvement of nearly 24% in the antenna's radiation efficiency following the incorporation of the graphite structure. The primary contributor to mutual coupling is the transition of surface currents from the first patch to the

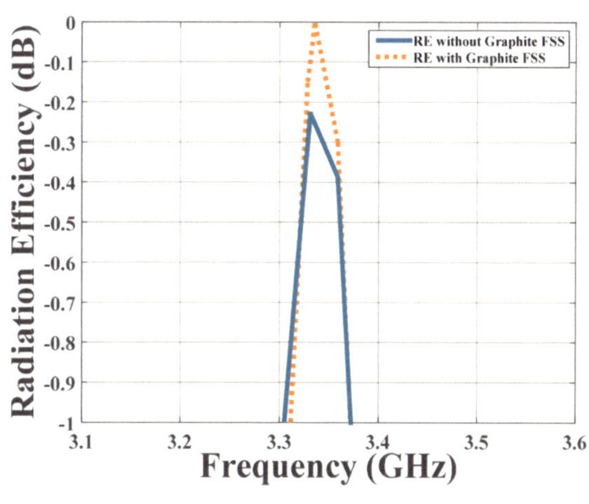

Figure 21.
Comparative diagram of radiation efficiency of the antenna without and with graphite structure.

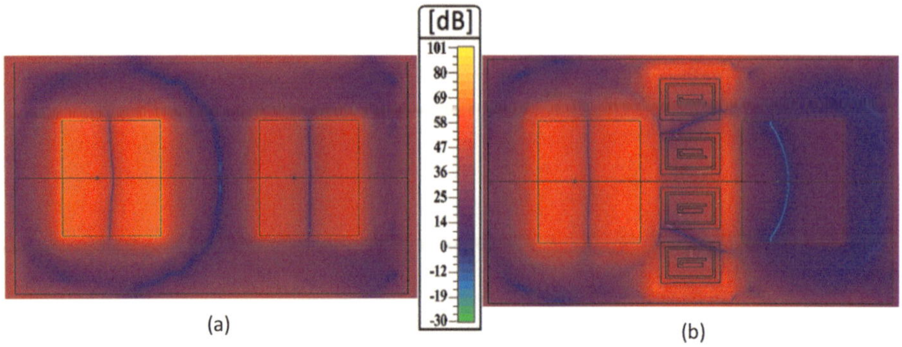

Figure 22.
Antenna surface current distribution: (a) without graphite structure and (b) with graphite structure.

second patch. **Figure 22** illustrates the distribution of surface currents in the antenna, showcasing the impact of the coupling reduction structure between the two radiating elements. Upon stimulating one of the patches, **Figure 22** demonstrates the designed structure's efficacy as a barrier, effectively impeding the transition of surface current from the first patch to the second patch. Further insight into the dynamics of electric fields is provided in **Figure 23**, where the graphite structure positioned between the two elements creates a vortex, effectively preventing the transition of electric fields from the first patch to the second patch. This visual representation emphasizes the role of the graphite structure in disrupting the flow of electric fields between the radiating elements, thereby contributing to the reduction of mutual coupling in the antenna system.

In the realm of communication channels, the transmission of electromagnetic waves encounters various obstacles, leading to signals reaching the receiver from diverse directions. Direct waves differ from those reflected in terms of amplitude and phase, introducing interference at the receiver, commonly referred to as "tach." Circular polarization antennas play a pivotal role in mitigating these challenges. When electromagnetic waves reflect, circular polarization antennas induce a change in the direction of polarization. For instance, if waves are initially upright, they turn left after reflection, and vice versa. This phenomenon significantly diminishes multidirectional interference. Unlike their linear polarization counterparts, circular polarization antennas do not require precise alignment for a reliable connection between the transmitter and receiver. Circular polarization antennas can send and receive signals with both horizontal and vertical components simultaneously. Consequently, the orientation of antennas in the receiver and transmitter becomes less critical. This research employs a structural approach designed to meet necessary specifications while ensuring simplicity and cost-effectiveness in the antenna manufacturing process. The utilization of circular polarization antennas aligns with the goal of streamlining mass production, making it both simple and economically feasible within the available facilities.

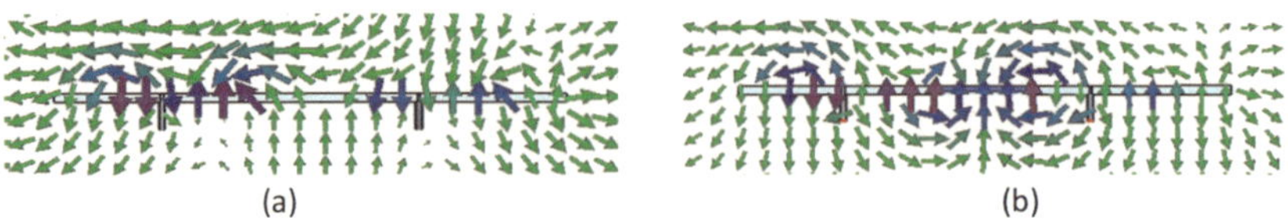

Figure 23.
Distribution of antenna electric fields: (a) without graphite structure and (b) with graphite structure.

4.3 Mitigating mutual coupling through DGS and EBG integration

In the realm of communication channels, obstacles abound, and electromagnetic waves encounter various directions before reaching the receiver. The waves directly received differ from those reaching the antenna via reflection, leading to disparities in amplitude and phase, causing interference known as "tach." Circular polarization antennas play a pivotal role in addressing these challenges. Reflection induces a change in the direction of polarization, significantly reducing multi-directional interference compared to linear polarization antennas. Unlike their linear counterparts, circular polarization antennas can send and receive both horizontal and vertical components, eliminating the need for precise alignment between the transmitter and receiver antennas. In this research, we introduce a structure designed to meet specifications while ensuring a straightforward and cost-effective antenna manufacturing process conducive to mass production within available facilities.

4.3.1 Description of the designed antenna

Circular polarization proves advantageous due to its ability to overcome the multi-way phenomenon, eliminate the need for intricate tuning of transmitter and receiver antennas, and facilitate frequency band reuse with left and right polarization. This property, crucial in various telecommunication systems, extends to military and industrial applications. To achieve circular polarization, a square patch with two cut corners is employed, fed by a probe through a coaxial line. The coaxial line power supply offers several advantages over micro-line alternatives, including lower power line losses, prevention of undesirable power line radiation, and the ability to adjust input impedance without requiring a quarter-wavelength line. The antenna patch is crafted on a 6035 RT microwave substrate with a dielectric coefficient (εr) of 3.6 and a 1 mm thickness, courtesy of the Rogers Company. Operating in the C-frequency band at approximately 6.39 GHz, the antenna is powered by a 50-ohm coaxial SMA probe. All dimensions adhere to standard formulas outlined in Refs. [21, 22]. **Figure 24** illustrates the structure of the designed

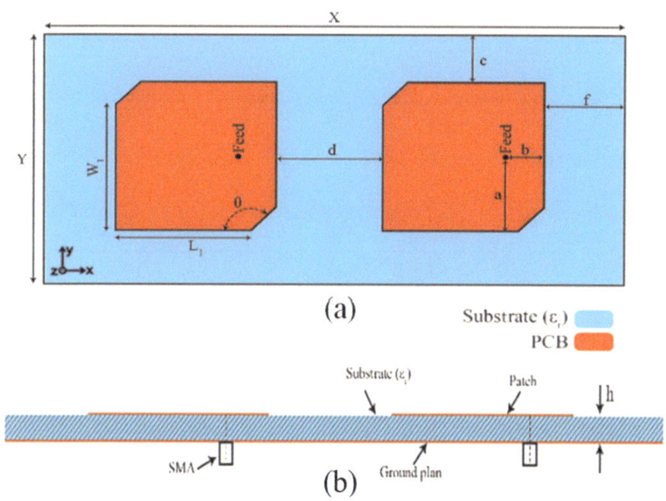

Figure 24.
Initially designed antenna view without placement of the coupling reduction structure: (a) top view and (b) side view.

Parameter	Dimensions (mm)	Parameter	Dimensions (mm)
X	70.7	L_1	10.6
Y	35.4	D	23.5
c	11.7	A	5.8
f	11.7	b	3.3
W_1	10.6	θ	135°
L_1	10.6	h	1
X	70.7	L_1	10.6

Table 6.
Values of designed antenna dimensions ***Figure 24.***

antenna, fed by a probe through the coaxial line, with radiant fields exhibiting right circular polarization. Altering the cut placement to the other two corners generates left circular polarization and **Table 6** shows the antenna dimension values.

In line with the resonator model, the square patch's predominant mode is 10TM. To induce circular polarization, cuts in the two corners along the patch's diameter are essential. Properly selecting square dimensions and cut amounts create two modes with equal amplitude and a 90-degree phase difference, intensifying at frequencies close to each other.

To comprehensively assess the impact of mutual coupling between two array antenna elements, a thorough analysis of the scattering parameters (S12) is imperative for both the pre and post application of the reduction structure. **Figure 25**, in the initial step, presents the schematic representation of the antenna's scattering parameters before implementing the coupling reduction structure, focusing on S11. Observing the S11 diagram in **Figure 25** reveals the antenna's resonant frequency to be 6.35 GHz, positioning it within the C-frequency band spanning 4 to 8 GHz. This frequency range holds significance as the primary allocation for Earth-satellite communication, particularly prevalent in commercial communication satellites. To delve into the polarization characteristics of the antenna, scrutiny of the axial ratio diagram becomes essential. **Figure 26** illustrates the axial ratio of the designed antenna

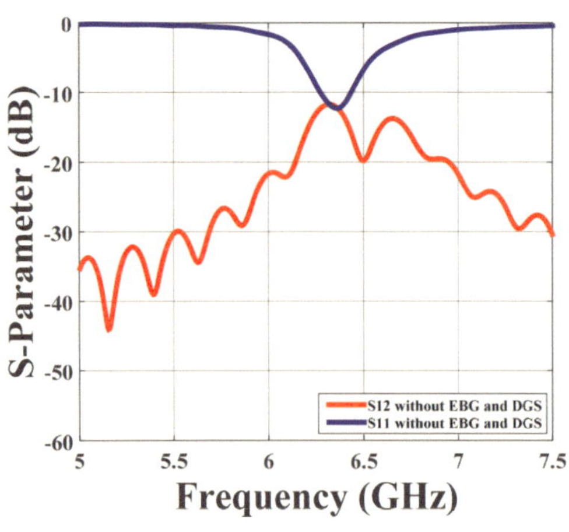

Figure 25.
Diagram of antenna scattering parameters without EBG and DGS structure.

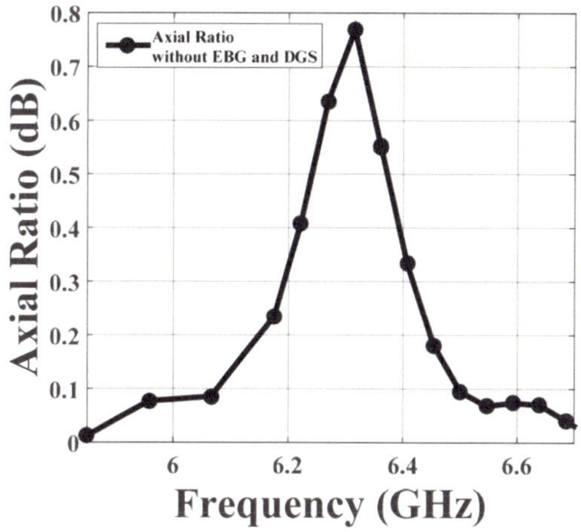

Figure 26.
Axial ratio diagram of the antenna without EBG and DGS structure.

concerning frequency. When the axial ratio value at the antenna's resonance frequency is below 3 dB, it signifies circular polarization, as depicted in **Figure 26**.

The antenna bandwidth, calculated using Eq. (7), is determined to be 0.5%:

$$\%FBW = \frac{f_{\text{high}} - f_{\text{low}}}{f_{\text{center}}} \tag{26}$$

The radiant pattern characteristics of the initially designed antenna are expounded in **Figure 27**. The simulated E-plane radiation pattern, showcased in **Figure 27**, captures the antenna's behavior at the resonant frequency of 6.35 GHz without the mutual reduction structure, attaining a maximum value of 10.4 dB.

4.3.2 Description of the designed structure for mutual coupling reduction

Electromagnetic bandgap (EBG) structures have emerged as a promising paradigm for innovative antenna designs, offering insights into dispersion diagrams, surface impedance, and reflection phase characteristics. EBG structures enable the

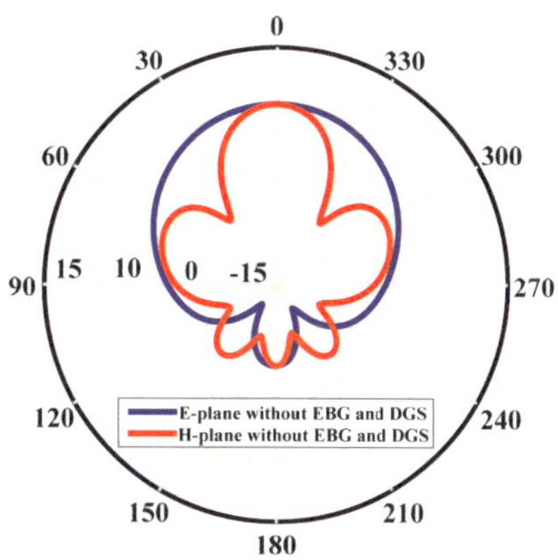

Figure 27.
Antenna radiation pattern diagram without EBG and DGS structure.

specification of stop-bands, pass-bands, and bandgaps. Essentially, an EBG structure comprises repeated elements of dielectric or metallic materials organized as a unit cell, strategically placed with specific repetitions. Conventionally, EBG structures find placement between antenna arrays, playing a crucial role in mitigating mutual coupling and enhancing reflected radiation [23, 24]. To achieve a targeted electromagnetic forbidden band, designers can introduce gaps in the ground plane with diverse shapes and configurations. This method employs a synergistic approach, integrating EBG and Defected Ground Structure (DGS) methods to craft a novel structure aimed at reducing mutual coupling. The EBG structure features multiple nested circles, while the DGS structure encompasses a section of the ground plate with two circular rings, as illustrated in **Figure 28**, showcasing the unit cell view of the designed structure. The structural analysis, including the extraction of the dispersion diagram to delineate the magnetic bandgap, utilizes the eigenmode method. **Figure 29** depicts the yellow

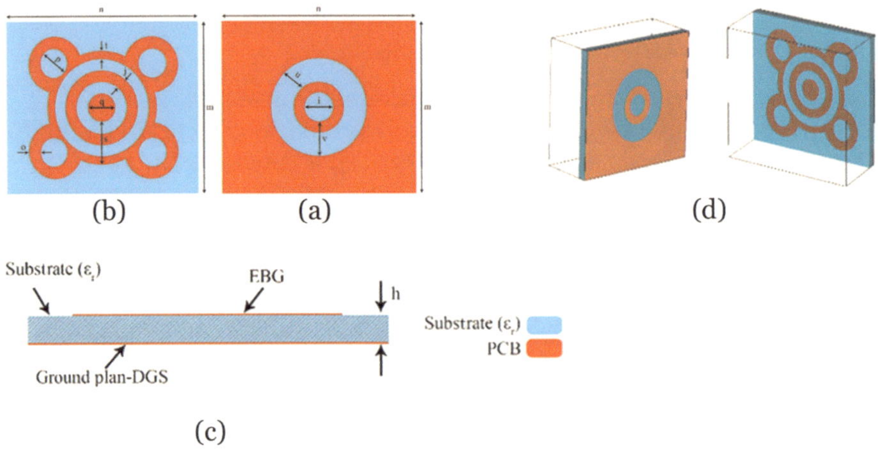

Figure 28.
View of the cell structure of the combined DGS and EBG unit to reduce mutual coupling, (a) DGS bottom view, (b) top view of EBG, (c) side view, and (d) three-dimensional view.

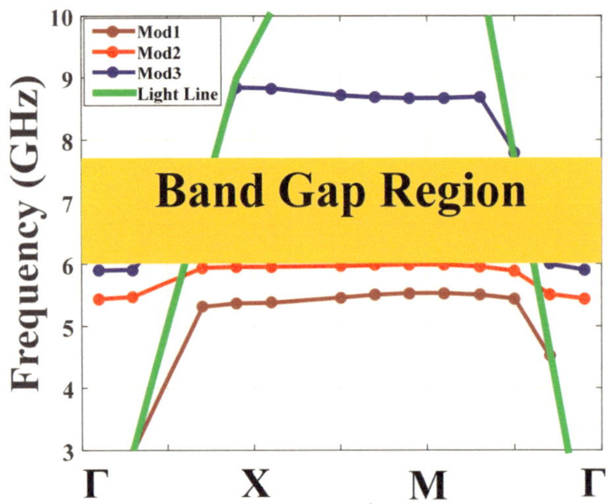

Figure 29.
Dispersion diagram output diagram in order to obtain the magnetic bandgap of the structure.

region symbolizing the magnetic bandgap of the structure, with the horizontal diagram denoted as ΓXMΓ. The interval [0,1] or [ΓX] correlates to the electrical phase spanning [0,180] degrees. Notably, the exclusion of one of the modes warrants consideration. This qualitative omission is addressed by scrutinizing the electric field vectors at a specific kx value (first Brillouin region (ΓX)) within the inner region of the light cone and the YZ plane, as determined by the structure's orientation. Given that this study explores the combined EBG and DGS structure for coupling reduction in flat array antennas, recognizing that electric field vectors inside the substrate align in the ±Z direction with a predominant vector direction (e.g., all vectors in the −Z direction), it becomes evident that the first mode deviates from this pattern and should be excluded from the bandgap calculation (**Figure 29**). **Table 7** also shows the dimensions of the designed structure in **Figure 28**.

The output dispersion diagram (**Figure 29**) reveals that the magnetic bandgap of the structure crafted in this investigation spans from 6 GHz to approximately 7.8 GHz. This signifies the presence of a magnetic band within this frequency range, indicating the exclusive application of coupling reduction during this specific span. Consequently, the designed structure remains inactive in frequencies outside this specified range. However, considering the antenna's operational frequency and acknowledging the inverse correlation between frequency and structural dimensions, diverse operational frequencies can be attained by modifying the structure's dimensions. By diminishing the structure's dimensions, a higher operational frequency can be achieved, whereas increasing the dimensions facilitates operation at a lower frequency. Consequently, it is imperative to recognize that a tailored design approach is necessary for each distinct frequency range. This nuanced strategy ensures optimal performance and targeted coupling reduction for specific operational frequencies.

4.3.3 Conclusive outcomes

In this segment, an examination of the mutual coupling between array antenna elements is undertaken, guided by the principles and configurations elucidated in Sections 2 and 3. The implementation of the synergistic EBG and DGS structure is a focal point in this analysis. **Figure 30** provides a visual representation of the antenna featuring coupling reduction structures positioned between its two radiating elements. This strategic integration is pivotal for assessing and quantifying the impact on mutual coupling within the array antenna system.

In this context, a comprehensive examination of the antenna's performance is presented through **Figures 25** and **31**, delineating the contrasting scenarios before and

Parameter	Dimensions (mm)	Parameter	Dimensions (mm)
n	14	s	3
m	12	o	0.8
t	0.6	u	1.6
j	0.8	i	2
p	2	v	2.4
q	2		

Table 7.
Unit cell values of ***Figure 28****.*

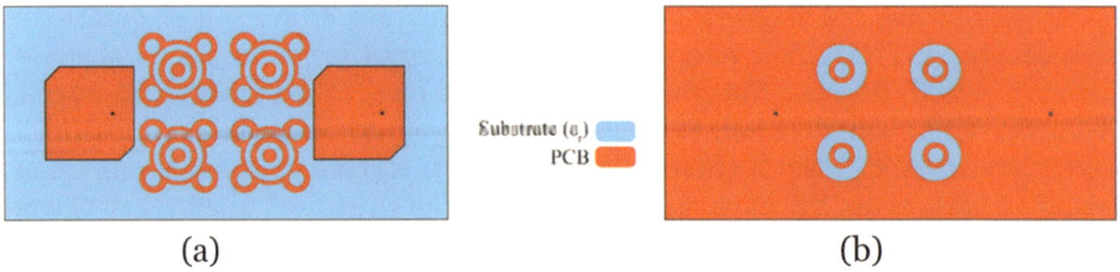

Figure 30.
Placement of the structure designed for mutual coupling, (a) underside view of the DGS structure antenna. (b) Upper view of EBG structure antenna.

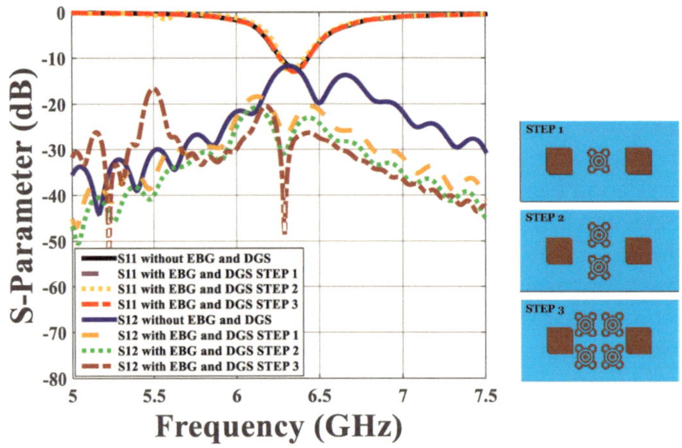

Figure 31.
Diagram of antenna dispersion parameters without and with EBG and DGS structure in three STEP with different unit cell arrangement.

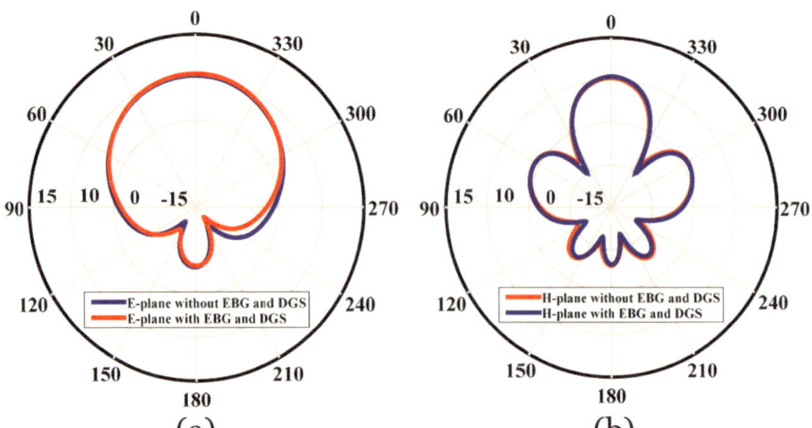

Figure 32.
Comparative diagram of unstructured antenna radiation pattern with EBG and DGS structure. (a) H-plane and (b) E-plane.

after implementing the devised structure aimed at mitigating mutual coupling among array antenna elements. **Figure 31** specifically explores the influence of varying the repetitions of the unit cell within the antenna patches, scrutinizing three distinct stages. The dispersion parameter S12 is instrumental in this analysis, revealing that the optimal configuration, denoted as STEP3 (involving 4 unit cell repetitions), achieves a remarkable 36.5 dB reduction in mutual coupling. **Figure 32** shows the comparison diagram of the radiation pattern. Additionally, **Figure 33** offers a comparative illustration of the axial ratio for the antenna, both in its original state and with the integrated EBG and DGS structure. This comparative assessment, as depicted in **Figure 33**, provides valuable insights into the impact of the devised structure on the antenna's axial ratio, offering a holistic view of its polarization characteristics in the presence of the coupling reduction strategy. Also, **Figure 34** shows the comparative diagram of antenna efficiency.

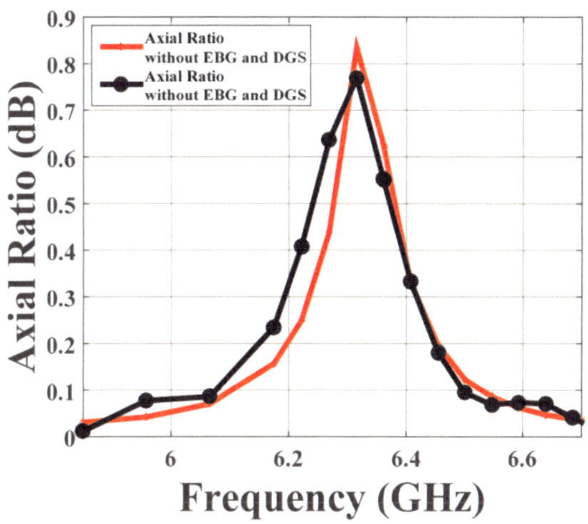

Figure 33.
Comparative diagram of axial ratio antenna without and with EBG and DGS structure.

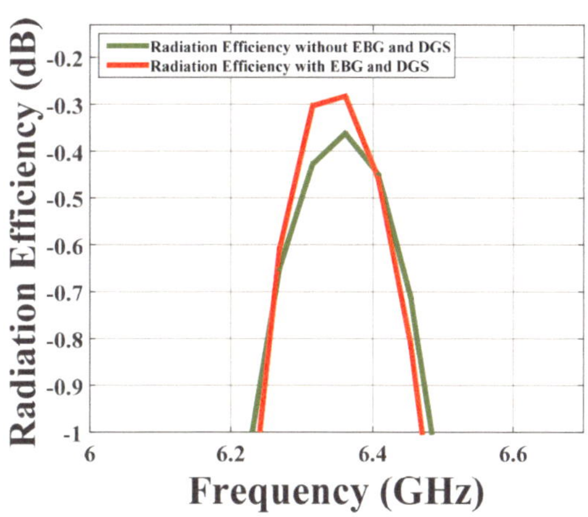

Figure 34.
Comparative diagram of radiation efficiency of antenna without and with EBG and DGS structure.

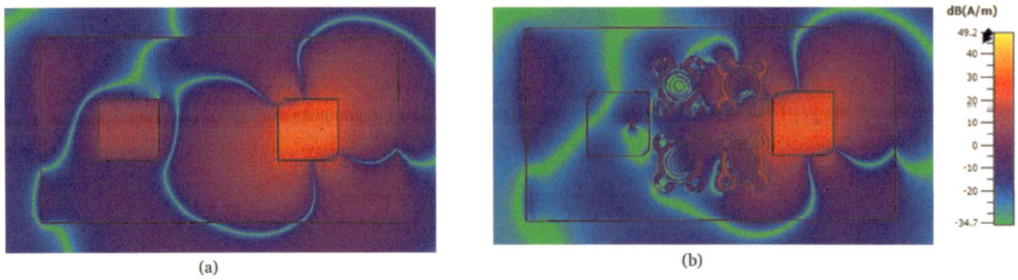

Figure 35.
Distribution of antenna surface currents (a) without EBG and DGS structure (b) with EBG and DGS structure.

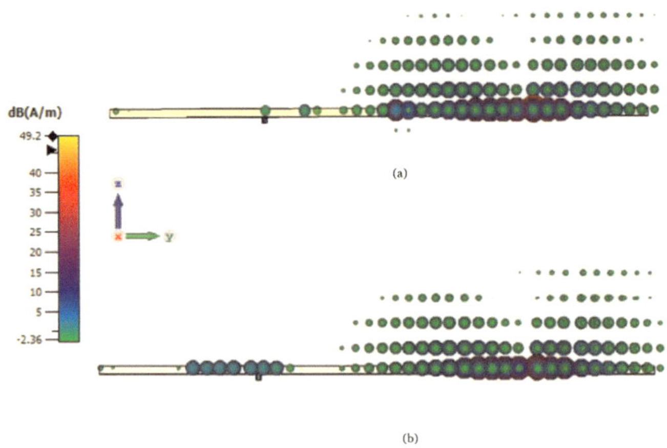

Figure 36.
Antenna magnetic fields (a) without EBG and DGS structure (b) with EBG and DGS structure.

4.3.4 Surface current distribution and magnetic field analysis

The primary determinant of mutual coupling lies in the transmission of surface currents from the first antenna patch to the second one. Illustrated in **Figure 35** is the surface current distribution in an antenna, both with and without a coupling reduction structure positioned between the two radiation elements. This visualization (**Figure 35**) elucidates how the devised structure, intended for mutual coupling reduction, functions as a barrier when one of the patches is stimulated, effectively impeding the transfer of surface current from the first patch to the second. Furthermore, **Figure 36** provides insight into the magnetic fields of the antenna, comparing scenarios with and without the coupling reduction structure between the two radiating elements. The magnetic field analysis offers a comprehensive view of the electromagnetic behavior, highlighting the efficacy of the coupling reduction structure.

5. Conclusion

In the first part of the book, an introduction to mutual coupling is explained, and in the second part, wireless interference is explained. In the third part, the theory of antenna cross-coupling reduction is investigated. In the third part, we have analyzed different methods of reducing mutual coupling. These methods include reduction of mutual coupling with liquid dielectric, reduction of mutual coupling with graphite materials, MTM, and EBG design, and reduction of mutual coupling through the integration of DGS and EBG, which was first proposed by Mohammadreza Rohaninezhad and his colleagues in 2023. and 202 were presented. Section 4.1 focuses on a specific application of water as a dielectric material to improve antenna performance, with aerial and naval applications in mind. For the first time, a method for lowering surface currents between the antenna array's two radiators has been proposed based on water. The liquid dielectric was inserted into hollow plexiglass containers. Pure water is used, and it has the ability to absorb and reradiate EM waves in partial proportions. To consider environmental variables, an ambient temperature of 4 to 40°C is assumed for both water and glass containment. EM waves from the first patch to the second patch collide with semiconductive water during coupling, which absorbs the waves. Using this method, the coupling was reduced by 15 decibels. The mutual coupling reduction in the extended operating frequency is one of the advantages of this method. According to the measurements, the coupling has been reduced in an extensive band. Furthermore, when compared to traditional methods, the amount of metal pieces used in this method has been reduced. The method presented here can be used in any antenna array, but it is especially useful for multidisciplinary communication. The proposed instrument is invaluable for integrating the antenna with solar panels for the mentioned application due to the optical transparency of the water and the contained material. Furthermore, the use of water rather than other dielectric materials allows for applications in naval communications. In Section 4.2, the application of graphite in reducing the mutual coupling of electromagnetic waves by combining the structure of EBG was investigated. This structure is one of the practical methods to reduce mutual coupling in array antennas. Considering that EBG and graphite substrate are complementary to each other, therefore, the graphite sheet is responsible for a part of reducing the mutual coupling, and the other part is responsible for the EBG structure. This structure can be applied to all antennas with a working frequency between 3.1 and 3.75 GHz, and there is no limit to its use. It was possible to further reduce mutual coupling by combining these two structures. Also, in the composition design, graphite should be placed in a place that reduces the overall weight of the antenna due to the lightness of graphite. This feature can significantly reduce the weight of the antenna in larger dimensions and more arrays. The results showed that the mutual coupling between the two elements is reduced by 30 dB, which in the MIMO system, the coupling reduction of more than 15 dB is a very good value. Also, the antenna's radiation efficiency was increased by nearly 24%, and the gain of the antenna was improved from 8.1 to 9.3 dB. By analyzing and also confirming the good performance of the structure in the antenna outputs, the basic parameters of the MIMO antenna, such as envelope correlation coefficient (ECC), diversity gain (DG), mean effective gain (MEG), total active reflection coefficient (TARC), and channel capacity loss (CCL), in the operating frequency band of the antenna showed that the obtained values show the positive effects of a graphite structure on antenna components. In Section 4.3, first, a circular array antenna was designed with the technique of cutting the corners of the patches. Then, by combining EBG and DGS methods, a single cell was designed to reduce mutual coupling. This is because, in previous studies, the combination of the two methods did not reduce the coupling in the antenna with circular polarization at a frequency of 6.3. Therefore, by applying a combination of EBG and DGS methods, the reduction of mutual coupling by 36.5 dB was calculated, which is an acceptable value compared to previous studies.

The advantages of this structure include improved antenna gain, antenna radiation efficiency, and antenna envelope correlation coefficient. This structure can be used in all transmitter and receiver systems to send waves from the ground to the satellite in telecommunication and military satellites in the C-frequency band. In less research, all antenna components have been improved together, and by designing this combined structure, the characteristics of all antenna components have been improved. This structure can be used in aperture array antennas and wherever there is a need to suppress surface waves in the frequency range of 6 to 7.8 GHz. Also, by changing the dimensions of the main elements of the designed unit cell, its bandgap frequency can be changed and this structure can be used in other frequencies.

Acknowledgements

The authors extend their sincere appreciation to the NAK Company Research Laboratory for providing the invaluable opportunity to carry out research in a real-world setting and for granting access to specific components for examination. The generous support from NAK Company has been instrumental in the successful completion of this study.

Conflict of interest

The authors declare no conflict of interest.

Author details

Mohammadreza Rohaninezhad*, Meysam Jalali Asdabadi, Changiz Ghobadi and Javad Nourinia
Department of Electrical Engineering, Urmia University, Urmia, West Azarbaijan, Iran

*Address all correspondence to: m.rohaninezhad@urmia.ac.ir

References

[1] Amitay N, Galindo V, Wu CP. Theory and Analysis of Phased Array Antennas. New York: Wiley-Interscience; 1972

[2] Hansen RC, editor. Microwave Scanning Antennas: Array Theory and Practice. New York and London: Academic Press; 1966

[3] Craeye C, González-Ovejero D. A review on array mutual coupling analysis. Radio Science. 2011;**46**(02): 1-25

[4] Hannan P. The element-gain paradox for a phased-array antenna. IEEE Transactions on Antennas and Propagation. 1964;**AP-12**:423-433

[5] Pozar D. A relation between the active input impedance and the active element pattern of a phased array. IEEE Transactions on Antennas and Propagation. 2003;**51**:2486-2489

[6] Rohaninezhad M, Ghayekhloo A, Afsahi M, Denidni TA. Design of a transparent system for mutual coupling reduction of microstrip array antennas with confined water. Physica Status Solidi (a). 2022;**219**(12): 2200082

[7] Rohaninezhad MR, Shirpay A, Soleimanyfar P. Design of an applied method with graphite structure for suppression of surface electromagnetic waves between array antennas. AIP Advances. 2023;**13**(4):045013

[8] Pakdin M, Ghayekhloo A, Rezaei P, Afsahi M. Transparent dual band Wi-Fi filter for double glazed energy saving window as a smart network. Microwave and Optical Technology Letters. 2019; **61**(11):2545-2550

[9] Gangwar D, et al. Characterization and performance measurement of low RCS wideband circularly polarized MIMO antenna for microwave sensing applications. IEEE Transactions on Instrumentation and Measurement. 2019;**69**(6): 3847-3854

[10] Ghayekhloo A, Afsahi M, Orouji AA, Denidni TA. Triangle and aperiodic metasurfaces for bistatic backscattering engineering. Physica Status Solidi (b). 2019;**256**(10):1900059

[11] Rohaninezhad M, Jalali Asadabadi M, Ghobadi C, Nourinia J. Design and fabrication of a super-wideband transparent antenna implanted on a solar cell substrate. Scientific Reports. 2023;**13**(1):9977

[12] Bogie IS. Conduction and magnetic signalling in the sea a background review. Radio and Electronic Engineer. 1972;**42**(10):447-452

[13] Li Q, Ding C, Yang R, Tan M, Wu G, Lei X, et al. Mutual coupling reduction between patch antennas using meander line. International Journal of Antennas and Propagation. 2018

[14] Louche F, Křivská A, Messiaen A, Wauters T. Simulation of cold magnetized plasmas with the 3D electromagnetic software CST Microwave Studio®. In EPJ Web of Conferences. EDP Sciences. 2017;**157**: 03031

[15] Afanasov IM, Morozov VA, Kepman AV, Ionov SG, Seleznev AN, Tendeloo GV, et al. Preparation, electrical and thermal properties of new exfoliated graphite-based composites. Carbon. 2009;**47**(1):263-270

[16] Cermak M, Perez N, Collins M, Bahrami M. Material properties and structure of natural graphite sheet. Scientific Reports. 2020;**10**(1).10672

[17] Qiu H, Wang J, Qi S, He Z, Fan X, Dong Y. Microwave absorbing properties of multi-walled carbon nanotubes/ polyaniline nanocomposites. Journal of Materials Science: Materials in Electronics. 2015;**26**:564-570

[18] Liu Q, Zheng Y, Zhong L, Cheng X. Removal of tetracycline from aqueous solution by a Fe3O4 incorporated PAN electrospun nanofiber mat. Journal of Environmental Sciences. 2015;**28**:29-36

[19] Jia K, Liu W, Li K, Wang D, Ma C. A novel synthesize approach and electromagnetic wave absorbing properties of expanded graphite/Fe3O4/ carbon nanorod composite microstructure. Journal of Materials Science: Materials in Electronics. 2019; **30**:17011-17019

[20] Wang C, Han X, Xu P, Zhang X, Du Y, Hu S, et al. The electromagnetic property of chemically reduced graphene oxide and its application as microwave absorbing material. Applied Physics Letters. 2011;**98**(7)

[21] Balanis CA. Antenna Theory: Analysis and Design. Hoboken, New Jersey, US: John Wiley & Sons; 2016. p. 1104

[22] Shirpay A, Rohaninezhad MR, Tavakoli M, Zarezadeh E. Reduction of mutual coupling in a microstrip array antenna with circular polarization in the C-frequency band using a combination of DGS and EBG methods. Engineering Research Express. 2023;**5**(1):015033

[23] Lindberg P, Morton L, Kaikkonen A, Cheng S, Hallbjorner P. Technique of ground size tuning for isolation between monopoles in compact wireless terminals. IEEE Transactions on Antennas and Propagation. 2012;**60**(11): 5488-5491

[24] Lu J, Kuai Z, Zhu X, Zhang N. A high-isolation dual-polarization microstrip patch antenna with quasi-cross-shaped coupling slot. IEEE Transactions on Antennas and Propagation. 2011;**59**(7):2713-2717